AQA BIOLOGY
Specification

*Further Studies in*

# HUMAN
# BIOLOGY

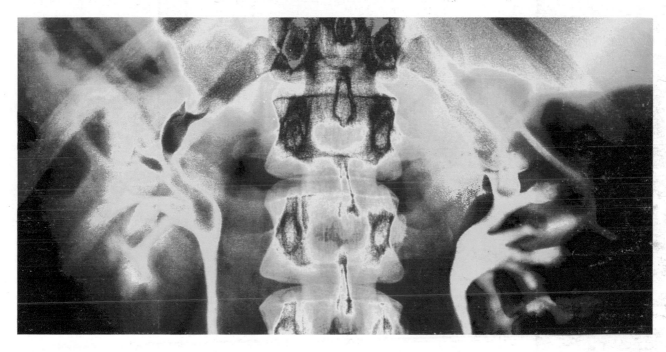

## Margaret Baker
## Bill Indge
## Martin Rowland

Hodder & Stoughton
A MEMBER OF THE HODDER HEADLINE GROUP

The Publishers would like to thank the authors of this book for providing the following photographs:

Introduction (whale) p4, 3.17 p69, 3.18 p69, 4.1 p73, 4.2 p73, 4.3 p74, 4.4 p75, 4.6 p76, 4.9 p79, 4.10 p81, 4.11 p81, 4.14 p83, 4.15 p84, 5.29 p108, 6.5 p117, 6.7 p117, 6.17 p124, 6.19 p125, 9.13 p193 and 10.9 p219.

**Action Plus Photographic:** 9.17 p197, 12.18 p261; **Bryan & Cherry Alexander:** 13.1a p265; **Allsport:** 8.1 p162; **Ardea:** 1.12 p15 (Steve Hopkin), 2.18 p47 (P. Morris), 3.16(b) p67 (Steve Hopkin), 3.16(c) p67 (Alan Weaving), 3.16(d) p67 (Åke Lindau), 3.16(f) p67 (Ken Lucas), 3.16(h) p67 (Jim Zipp), 5.20 p103 (Ian Beames), 6.1 p113 (Anthony & Elizabeth Bomford); **Anthony Blake Photo Library:** 3.2 p54 (Gerrit Buntrock); **Bubbles:** 8.3 p164 (Lucy Tizard), 8.17, p179 (Frans Rombout), 9.21 p203 (Katie Van Dyck); **John Cleare/Mountain Camera Picture Library:** 10.1 p214; **Bruce Coleman Collection:** 2.8 p37; **Collections:** 6.6 p117 (Ray Farrar); **Frank Lane Picture Agency:** 2.16 p46 (David Hosking); **Garden Matters:** 3.13(d,e) p65; **Holt Studios International:** 3.3t p55 (Sarah Rowland), 3.3b p55 (Nigel Cattlin), 6.11 p120, 6.12 p121; **Roger G Howard:** 8.15 p175; **International Artists Ltd:** 1.1 p7; **Nigel Luckhurst:** 9.5 p186; **Natural History Photographic Agency:** 1.21 p27, 2.11 p43 (Daniel Heuclin), 2.15 p46 (Stephen Dalton), 3.1 p54 (Kevin Schafer), 3.7(d) p62 (J&M Bain), 3.9 p63 (Stephen Dalton), 3.13(a,b) p65 (Stephen Dalton), 3.13(c) p65 (GJ Cambridge), 3.16(g,i) p67 (John Shaw), 5.19 p102 (Stephen Dalton); **Oxford Scientific Films:** 2.1 p31 (Doug Allan), 3.16(a) p67 (Mark Deeble & Victoria Stone), 3.16(e) p67 (Zig Leszczynski), 5.24 p105 (Harold Taylor), 8.9 p170 Ronald Toms; **PA Photos:** 7.1 p 132; **Panos Pictures:** 13.1b p265 (Neil Cooper); **Redferns:** 8.14 p174 (Ebet Roberts); **Science Photo Library:** 1.2 p7 (Dept of Clinical Cytogenetics, Addenbrooke's Hospital); 1.3 p8 (L Willatt, East Anglian Regional Genetics Services), 1.7 p10 (Science Pictures Ltd), 2.19 p47 (Eric Grave), 3.6 p61 Institut Pasteur, 3.7(a) p62 (Eye of Science), 3.7(b) p62 (Claude Nuridsang & Marie Perennou), 3.7(c) p62 (Manfred Kage), 5.6 p96 (Dr Jeremy Burgess), 7.7 p135 (Secchi Lecaque/CNRI), 7.11 p137, 7.17b p141 (D. Phillips), 8.2 p164 (Alex Bartel), 9.1 p182, 9.8(b) p189 (Astrid & Hanns-Frieder Michler), 9.8(c) p189 (Marshall Sklar), 9.8(d) p189 (Quest), 10.3 p215, 10.6 p218 (Francis Leroy), 10.8 p219 Dr Gopal Murti, 12.3 p247, 12.11 p253 (Biology Media), 13.2 p265 (Biophoto Associates), 13.12 p272 (Mark Clarke), 13.15 p274 (Dr Raymond P Clark, Clinical Research Centre, Middlesex & Mervyn Goff AIIP, LRPS Moorfield Eye Hospital, London); **Topham Picturepoint:** 8.9 p170 (Press Association), 9.19 p201, 9.24 p206.

Orders: please contact Bookpoint Ltd, 130 Milton Park, Abingdon, Oxon OX14 4SB. Telephone: (44) 01235 827720, Fax: (44) 01235 400454. Lines are open from 9.00–6.00, Monday to Saturday, with a 24 hour message answering service. Email address: orders@bookpoint.co.uk

*British Library Cataloguing in Publication Data*
A catalogue record for this title is available from The British Library

ISBN 0 340 802456

First published 2001
Impression number   10  9  8  7  6  5  4  3  2  1
Year                        2007  2006  2005  2004  2003  2002  2001

Copyright © 2001 Margaret Baker, Bill Indge, Martin Rowland

Typeset by Cambridge Publishing Management Ltd

Printed in Italy for Hodder & Stoughton Educational, a division of Hodder Headline Ltd, 338 Euston Road, London NW1 3BH.

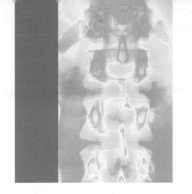

# Contents

|  | Introduction | 4 |
| Unit 1 | Transmission of Genetic Information | 7 |
| Unit 2 | Variation and Selection | 31 |
| Unit 3 | The Evolution and Classification of Species | 54 |
| Unit 4 | Numbers and Diversity | 73 |
| Unit 5 | Energy transfer | 93 |
| Unit 6 | Decomposition and Recycling | 113 |
| Unit 7 | Reproduction and Development | 132 |
| Unit 8 | Growth and Development to Maturity | 162 |
| Unit 9 | Digestion and Diet | 182 |
| Unit 10 | Transport of Respiratory Gases | 214 |
| Unit 11 | Nervous System | 226 |
| Unit 12 | Receptors and Effectors | 246 |
| Unit 13 | Homeostasis | 265 |
|  | Index | 286 |

# Introduction

The photograph shows a southern right whale. It is a marine mammal and can be seen off the coasts of Australia, South Africa and South America. There is a northern right whale as well. It is found in the northern Atlantic and Pacific oceans. These whales are different species. They are similar in appearance because it is thought that they share a common ancestor, but there are also slight differences between them which have arisen as the result of the two species being isolated from each other for thousands of years. The first three chapters in this book look at genetics and evolution and should help you to understand how species such as the southern right whale evolved.

A southern right whale surfacing after being under water for 30 minutes

If you look at the photograph carefully you will see number of lumps on the whale's head. These are structures called callosities and they have a life of their own! They are covered with small animals called barnacles. Barnacles feed on tiny photosynthetic organisms which they filter out from the surrounding water. Whale lice and parasitic worms live among the barnacles. The relationships between the whale and the animals that live on its callosities form part of its ecology. Ecology is the study of the relationships between organisms and their environment and we shall look at this aspect of biology in chapters 4–6.

A whale can remain underwater for a long time. In southern right whales, dives may last for over 30 minutes. These animals have many adaptations which allow them to do this. Their gas exchange and blood systems, for example keep vital organs such as the brain supplied with oxygen while the whale is underwater. The way in which different systems of the human body function throughout life is the subject of the remaining chapters in this book.

So, if you are a human biologist and are concerned about conserving the southern right whale, or any other organism, you will need to study its genetics, its ecology and its physiology, and understand how all these different aspects are linked together. Although this book is divided into chapters and each chapter concentrates on a particular topic, it is important to appreciate that all these areas of biology are related to each other. We have tried to help you to understand the relationships between topics by using text questions, extension boxes and the assignments which are found at the end of individual chapters.

This book has been written in a similar way to the AS textbook *A New Introduction to Human Biology*. Each chapter shares a number of features.

## The chapter opening

Chapters start with a topical introduction. In most cases, this is an interesting application or an unusual aspect of biology related to the content of the chapter.

## The text

Your AS course should have provided you with a sound understanding of many fundamental biological ideas. The material in this book builds on these concepts and develops many of them further. In writing each chapter we have again tried to help you to gain a good understanding of basic principles rather than providing you with a lot of unnecessary detail and too much technical language. The book has been illustrated in colour throughout. The drawings, photographs and information contained in the captions should provide you with additional help in gaining an understanding of the subject.

## Text questions

The text questions should help you to understand what you have just been reading. They are meant to be answered as you go along. Many of them are straightforward and can be attempted from the information in the preceding paragraphs. Some of them, however, require you to apply this knowledge to a new situation or link it with material covered earlier in your course. We have not included any answers to these questions. If you get stuck, try reading the previous paragraph again. If you still have difficulties make a note of the question and get some help.

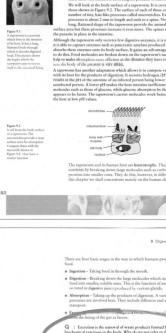

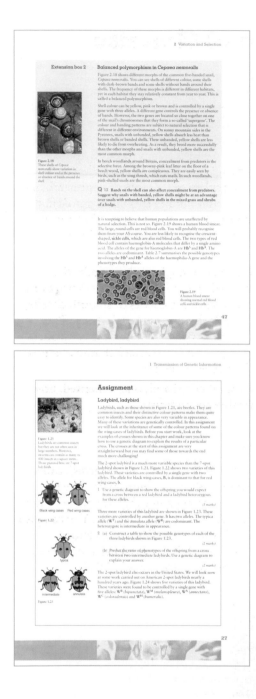

## Extension boxes

The extension boxes have several functions. Where we have a worked example of a calculation or a problem we have used a box to separate it from the main text. This should make the chapter easier to read. Extension boxes will also tell you more about some topics and help you to appreciate some of the links between different aspects of the subject.

## Summary

At the end of each chapter there is a summary. You can use it as a checklist to make sure that you have done everything your specification requires. It summarises what you need to know when you have finished a particular topic.

## Examination questions

The examination questions are reproduced or adapted by permission of the Assessment and Qualifications Alliance (AQA). It should be noted that these questions are based on the previous syllabus and are only intended to give you an idea of what might be asked.

## Assignment

Studying human biology entails a lot more than learning facts. It involves acquiring a range of skills. As a biologist you should be able to apply knowledge to new situations; interpret drawings and photographs, graphs and tables; and show an understanding of how the different aspects of the subject link together. In addition, you should be able to use mathematical skills to carry out a range of calculations and you must be able to communicate your knowledge and ideas effectively using appropriate biological language. If you want to get high grades in your Unit tests you will have to learn the necessary facts but you will also need a range of other important skills. The purpose of the text in this book is to provide you with the necessary understanding of the facts that you will need; the purpose of the assignments is to help you to master the skills you require.

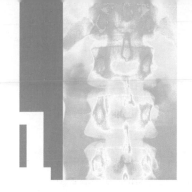

# Transmission of Genetic Information

**Figure 1.1**
The comedian Paul Merton has attached ear lobes. Whether your ear lobes are attached to the side of your head or not is controlled by a single gene

The comedian Paul Merton (Figure 1.1) has them. Do you? Check the fleshy lobes of your ears. Are they attached to the sides of your head, like Paul's, or are they unattached so that you can flip them backwards and forwards? This characteristic is controlled by a single gene that has two forms, called alleles. The allele that causes unattached ear lobes is dominant over the allele that causes attached ear lobes.

Other facial features that are controlled by single genes include: the presence or absence of freckles; a hairline that forms a widow's peak or is continuous across the forehead; dimples in the cheeks or smooth cheeks; a cleft chin or a smooth chin; long eyelashes or short eyelashes; a Roman nose or a straight nose. In each case, the first characteristic is controlled by a dominant allele of the relevant gene and the second characteristic is controlled by a recessive allele of the relevant gene.

This chapter explains how genes are inherited and, once inherited, how they interact to produce an observable characteristic.

## Chromosomes and cell division

Figure 1.2 shows a single human chromosome. It has two 'arms', called **chromatids**, held together by a region, called the **centromere**. The two chromatids are identical copies of the chromosome that are made by the semi-conservative replication of DNA, prior to cell division. You can also see dark and light bands along the chromosome. These are stained regions of DNA that are 6 to 10 Mb long (1 Mb represents a sequence of 1 megabase, i.e. 1000 organic bases in the nucleotides of DNA).

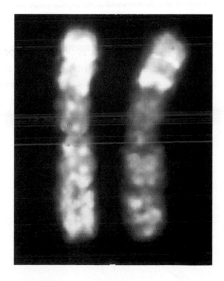

**Figure 1.2**
A pair of human chromosomes, showing the dark and light bands that result from specific staining. These bands represent similarities at the 6–10 Mb level of chromosome structure

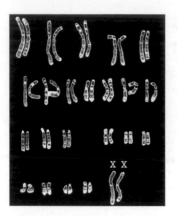

**Figure 1.3**
The chromosomes in a human female cell. The chromosomes in each pair are called homologous chromosomes. They contain genes controlling the same characteristics arranged in the same sequence

**Figure 1.4**
Human egg and sperm cells are haploid. At fertilisation a diploid cell is formed

Figure 1.3 shows all the chromosomes in a female human cell. The biologist who made this illustration first took a photograph of a dividing body cell. She then cut each chromosome from the photograph and arranged the cut chromosomes in descending order of length. You can see that:

● there are 46 chromosomes. With some exceptions, which we will examine later, this is the usual number of chromosomes in human cells.

● the chromosomes have been arranged as 23 pairs. The members of each pair are called **homologous chromosomes**. They not only look the same – they contain genes controlling the same characteristics, arranged in the same sequence.

**Q** 1 **Suggest why the chromosomes in each pair in Figure 1.2 have an identical pattern of dark and light bands.**

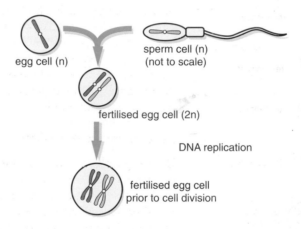

We need to understand why cells, such as the one shown in Figure 1.3, have pairs of homologous chromosomes. Figure 1.4 shows what happens when a human egg cell is fertilised by a human sperm cell. To make the drawing easier to follow, only one pair of homologous chromosomes is represented. The egg cell and sperm cell each have only one chromosome from each homologous pair: cells like this are called **haploid** and their chromosome number is represented as (n). When the sperm cell fertilises the egg cell, its chromosome is incorporated into the nucleus of the fertilised egg. This cell now has a pair of homologous chromosomes. Cells like this are called **diploid** and their chromosome number is represented as (2n).

During your AS course, you learned how cells replicate their chromosomes and separate them during mitosis forming new cells with the same chromosome number as the parent cell (i.e. diploid parent cells → diploid daughter cells or haploid parent cells → haploid daughter cells). Eggs and sperms are haploid cells formed by the division of diploid parent cells. The cell division involved is meiosis, which we will now study in more detail.

**Figure 1.5**
From a diploid parent cell, mitosis produces genetically identical diploid daughter cells whereas meiosis produces haploid cells that are genetically different from each other. To make the diagrams simple, only one pair of homologous chromosomes has been shown

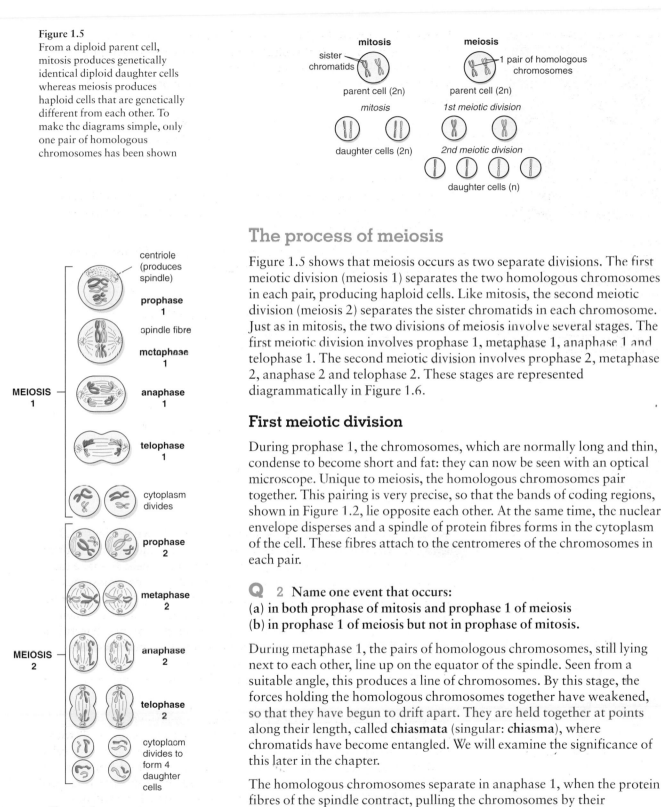

Figure 1.6
The behaviour of chromosomes during meiosis

## The process of meiosis

Figure 1.5 shows that meiosis occurs as two separate divisions. The first meiotic division (meiosis 1) separates the two homologous chromosomes in each pair, producing haploid cells. Like mitosis, the second meiotic division (meiosis 2) separates the sister chromatids in each chromosome. Just as in mitosis, the two divisions of meiosis involve several stages. The first meiotic division involves prophase 1, metaphase 1, anaphase 1 and telophase 1. The second meiotic division involves prophase 2, metaphase 2, anaphase 2 and telophase 2. These stages are represented diagrammatically in Figure 1.6.

### First meiotic division

During prophase 1, the chromosomes, which are normally long and thin, condense to become short and fat: they can now be seen with an optical microscope. Unique to meiosis, the homologous chromosomes pair together. This pairing is very precise, so that the bands of coding regions, shown in Figure 1.2, lie opposite each other. At the same time, the nuclear envelope disperses and a spindle of protein fibres forms in the cytoplasm of the cell. These fibres attach to the centromeres of the chromosomes in each pair.

**Q** 2 Name one event that occurs:
(a) in both prophase of mitosis and prophase 1 of meiosis
(b) in prophase 1 of meiosis but not in prophase of mitosis.

During metaphase 1, the pairs of homologous chromosomes, still lying next to each other, line up on the equator of the spindle. Seen from a suitable angle, this produces a line of chromosomes. By this stage, the forces holding the homologous chromosomes together have weakened, so that they have begun to drift apart. They are held together at points along their length, called **chiasmata** (singular: **chiasma**), where chromatids have become entangled. We will examine the significance of this later in the chapter.

The homologous chromosomes separate in anaphase 1, when the protein fibres of the spindle contract, pulling the chromosomes by their centromeres to opposite poles of the cell. In anaphase 1 each chromosome still has its sister chromatids together.

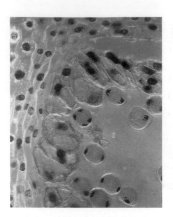

**Figure 1.7**
Most of the cells here are at Anaphase I with chromosomes at the spindle poles. The cell divides and the nuclear membrane reforms (Telophase I). These stages are repeated (Prophase II–Telophase II) to produce 4 haploid cells. At centre is a cell at Telophase II

During telophase 1, the chromosomes bunch up together, the cytoplasm of the cell begins to divide into two and the spindle disintegrates. Sometimes, after this, the chromosomes become long and thin again but usually the second meiotic division occurs straight away.

## Second meiotic division

At the start of the second division, the two daughter cells from meiosis 1 have one chromosome from each homologous pair, i.e. they are haploid. During prophase 2, a spindle forms in each cell at right angles to the now disintegrated spindle from prophase 1. If the chromosomes became long and thin at the end of meiosis 1, they now condense to become short and fat again.

During metaphase 2, the chromosomes in each cell line up on the equator of the new spindle. In metaphase 2 you can see the lines of chromosomes, each made of two sister chromatids, on the equators of the spindles. After a while, the centromeres holding the sister chromatids together divide and contraction of the spindle fibres pulls sister chromatids to opposite poles of the spindles.

At the poles of the spindles, the chromosomes bunch up into a tight ball and a nuclear envelope forms around each new nucleus. This is telophase 2 and marks the end of meiosis. The two divisions have produced four daughter cells, each of which contains only one chromosome from each homologous pair, i.e. is haploid.

**Q** 3 During which of the two meiotic divisions are homologous chromosomes separated?

**Extension box 1**

## When homologous chromosomes fail to separate

Very occasionally, the homologous chromosomes of one pair are pulled to the same pole of the spindle during anaphase 1 of meiosis. This is known as **chromosome non-disjunction** and results in one daughter cell with two copies of the chromosome and one daughter cell with no copies of the chromosome. If the gamete that contains both homologous chromosomes of a pair is fertilised by a gamete that also has one copy of that chromosome, we end up with a zygote that has three copies of the same chromosome, i.e. is trisomic for that chromosome.

Non-disjunction of the **X** chromosomes in a human female can lead to eggs that have two copies of the **X** chromosome (**XX**) or eggs that lack an **X** chromosome (designated **O**). If the **XX** egg cell is fertilised by an **X**-carrying sperm, an **XXX** zygote is formed. This develops into a phenotypically normal female. The effects are more serious if the **XX** egg cell is fertilised by a **Y**-carrying sperm. The resulting **XXY** zygote develops into a sterile male, a condition known as **Klinefelter's syndrome** that occurs in about 1 in every 500 live male births in the UK.

If the egg cell lacking an **X** chromosome is fertilised by a **Y**-carrying sperm, the resulting **OY** zygote fails to develop. Apparently, humans cannot survive without the genes on the **X** chromosome. If the **O** egg cell is fertilised by an **X**-carrying sperm, an **XO** zygote is produced that develops into a female who shows stunted physical growth and whose sex organs do not mature. This is called Turner's syndrome and occurs in about 1 in every 2000 live female births in the UK.

Chromosome non-disjunction can also occur with **autosomes** (chromosomes other than the sex chromosomes). A case with which you are likely to be familiar involves non-disjunction of chromosome 21 during the production of human eggs. After fertilisation, a zygote with three copies of chromosome 21 is produced. This zygote develops into a person who has Down's syndrome.

## Meiosis and genetic variation

During your AS course, you saw that mitosis produces cells that are genetically identical to each other and to the parent cell. This is important during growth and cell repair. Any organism that reproduces by mitosis will produce offspring that are genetically identical to itself and to each other. This is not the case when meiosis is involved. Two events that occur during meiosis result in genetic differences between the resulting daughter cells: independent assortment of homologous chromosomes and crossing over.

### Independent assortment of homologous chromosomes

If you look back to Figure 1.4, you can see that, in a human cell, the homologous chromosomes in each pair were derived from either the egg cell or the sperm cell. Although the homologous chromosomes in each pair have genes controlling the same characteristics in the same sequence, those genes do not necessarily have the same sequence of bases. Different versions of the same gene are called **alleles**, which we will consider later in this chapter. The point to realise here is that the two homologous chromosomes of a pair are not genetically identical to each other.

So far, we have considered only one of the pairs of homologous chromosomes in a cell. We have seen that the homologous chromosomes are separated during anaphase 1. The pole to which each chromosome from the pair is pulled is completely random. For example, the chromosome being pulled to the left pole of the cell during anaphase 1 in Figure 1.6 would be just as likely to have been pulled to the right pole: whichever pole it had been pulled to, its homologous partner would have been pulled to the opposite pole.

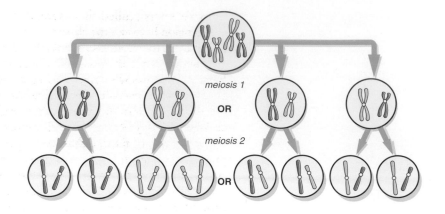

**Figure 1.8**
Independent assortment of chromosomes during meiosis results in genetically different daughter cells

Figure 1.8 shows two pairs of homologous chromosomes in a cell: those in red were derived from the egg cell and those in blue were derived from the sperm cell. The chromosomes in each pair are pulled at random to one of the poles of the spindle during anaphase 1. The movement of the chromosomes in one pair has no effect on the direction in which chromosomes of the second pair are pulled, i.e. they are pulled independently of each other. This shows independent assortment of homologous chromosomes. As a result of **independent assortment**, parent cells like those in Figure 1.8 could produce four kinds of genetically different daughter cells. As a general rule, independent assortment of chromosomes in a diploid cell (2n chromosomes) can result in any of $2^n$ different possible combinations of chromosomes in the four haploid cells produced by meiosis.

**Q** 4 **A human male has 23 pairs of chromosomes. How many different combinations of these chromosomes are possible in the sperm he produces?**

## Chiasmata and crossing over

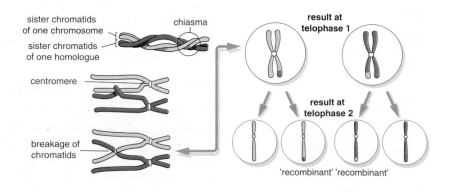

**Figure 1.9**
A chiasma shows where chromatids become entangled during prophase 1 of meiosis. Often the chromatids break at these points. When the broken fragments join the 'wrong' chromosome, genetic crossing over takes place

So far, we have assumed that chromatids never break. As they condense in prophase 1 of meiosis, the paired arms of the sister chromatids in each pair of homologous chromosomes often become intertwined to form one or more 'knots', called chiasmata. A single **chiasma** is shown in Figure 1.9. The tension between the chromatids at the point of a chiasma causes the chromatids to break. Usually the breakages are mended but the broken fragments do not always rejoin the chromatid from which they came. Figure 1.9 shows what happens if a chiasma occurs between the chromatids of one pair of homologous chromosomes and the broken pieces are joined to the wrong chromatid. Two of the resulting daughter cells have a copy of one of the chromosomes found in the parent cell. However, two of the daughter cells have a chromosome that was not present in the parent cell: it contains some genes from a chromatid in one of the homologous chromosomes but some genes from a chromatid of the other homologous chromosome. Because genes on these fragments have crossed from one homologous chromosome to another, we say that **crossing over** of genetic material has occurred.

**Q** 5 **What effect would crossing over have on the possible number of genetically different daughter cells that could be produced by the parent cell in Figure 1.8?**

## The principles of Mendelian inheritance

Gregor Mendel was an Austrian monk whose experiments with garden peas laid the foundations of our understanding of inheritance.

### Phenotype and genotype

We will consider the inheritance of a number of characteristics, such as wing length in fruit flies and flower colour in snapdragons. Each of these characteristics is referred to as a phenotype. Although you could list many **phenotypic** features for any organism, you will not be expected to solve problems with more than two phenotypic characteristics in your A2 Unit test.

Each phenotypic characteristic of an organism results from the interaction of two factors: the effect of the environment on the development of the organism and the effect of the combination of genes in the organism's chromosomes. The combination of genes that affects the inheritance of a phenotypic feature is called the **genotype**.

### Inheritance of sex in humans

Figure 1.3 shows the chromosomes of a human female. One of the pairs of homologous chromosomes is labelled **X**: these are the sex chromosomes. Like the other homologous chromosomes in each pair shown in Figure 1.3, the two **X** chromosomes have the same shape and size. Human males do not have two **X** chromosomes. Instead, they have one **X** chromosome and a much shorter chromosome, called a **Y** chromosome (Figure 1.10). We can represent a human female as **XX**, meaning that each of her body cells has two **X** chromosomes. A human male is **XY**, each of his body cells has one **X** chromosome and one **Y** chromosome.

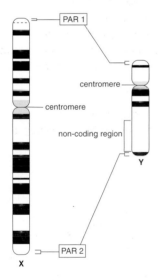

**Figure 1.10**
The sex chromosomes of a human male. The **X** chromosome contains a large number of genes, most of which are not concerned with sex determination. The shorter **Y** chromosome contains only a small number of genes, most of which affect the development of testes. Only the PAR 1 and PAR 2 regions pair with each other at meiosis

**Figure 1.11**
The inheritance of sex in humans. This pattern of inheritance is unusual in that it involves whole chromosomes, rather than individual genes

| parental phenotypes | female | x | male |
|---|---|---|---|
| parental genotypes | **XX** | | **XY** |
| gametes | (X) | | (X) (Y) |
| offspring (1) genotypes | **XX** | or | **XY** |
| offspring (1) phenotypes | female | | male |
| expected ratio of phenotypes | 1 | : | 1 |
| expected frequency of phenotypes | 0.5 | | 0.5 |

We can now examine what happens to the **X** and **Y** chromosomes when two humans have a child. This is represented in Figure 1.11. The notation used in Figure 1.11 is one that we will use to represent all genetic crosses and that you should use in a Unit test. In working through genetic crosses, it is important to remember that:

● parent cells are diploid

● gametes are haploid

● the zygote, produced by fertilisation, is diploid.

Figure 1.11 shows that all the eggs a human female will produce are identical, i.e. they contain one **X** chromosome. For this reason, human females are known as the **homogametic sex**. In contrast, human males produce two types of sperm: those with an **X** chromosome and those with a **Y** chromosome. The male is the **heterogametic sex**. If the egg is fertilised by a sperm carrying an **X** chromosome, the zygote will be **XX** and will be female. If the egg is fertilised by a sperm carrying a **Y** chromosome, the zygote will be **XY** and will be male.

**Q** 6 Explain the terms:
(a) **haploid**
(b) **homogametic**.

In theory, a male produces equal numbers of sperm carrying the **X** and **Y** chromosomes. Consequently, there is an equal chance of the egg being fertilised by an **X**-carrying or **Y**-carrying sperm. This enables us to calculate the expected frequency of the possible phenotypes of the offspring. Figure 1.11 shows this frequency as a decimal value: the expected frequency of a female is 0.5 and the expected frequency of a male is 0.5. We could express this in different ways, for example there is a 1 in 2 chance of a female, there is a 50% chance of a female, there is a 50:50 chance of a female or there is a 1:1 ratio of females to males. We will return to this idea of expected frequencies later in the chapter.

The inheritance of sex is unusual in that it involves whole chromosomes. In the other examples of inheritance that follow, we will be concerned not with entire chromosomes but with individual genes: we will usually ignore the chromosome that each gene lies on. Each gene controls the production of one polypeptide and is found at a specific place on one chromosome, called the **locus** of that gene.

**Figure 1.12**
Fruit flies of the species
*Drosophila melanogaster*. Some
of these flies have long wings
and some have vestigial wings.
This phenotype is controlled by
a single gene

## Monohybrid inheritance with dominance

In **monohybrid inheritance**, we consider the effects of a single gene.
Figure 1.12 shows flies belonging to the species *Drosophila
melanogaster*. Because they are easy to keep in the laboratory and have a
relatively short life cycle, *D. melanogaster* have been used extensively to
study inheritance. In the UK, *D. melanogaster* can be found in summer
around ripe, or rotting fruit, hence their common name of fruit flies.

The fruit flies in Figure 1.12 show variation in one phenotypic
characteristic: some have long wings and some have short (vestigial)
wings. This characteristic is controlled by a single gene that is located at a
locus on one of the chromosomes in the cells of *D. melanogaster*. There
are two varieties of this gene, called **alleles**. One allele of the gene for
wing length (represented as L) carries a base sequence that results in long
wings, the other allele of the gene for wing length (represented as l)
carries a base sequence that can result in vestigial wings. Table 1.1 shows
the possible genotypes involving these alleles and the phenotypes that
result. Notice that the l allele has no effect unless it is in the homozygous
form. This allele is called **recessive** and the L allele, that always shows its
effect if present in the genotype, is called the **dominant** allele.

| Genotype | Description of genotype | Phenotype |
|---|---|---|
| LL | Homozygous L (Homozygous dominant) | Long wings |
| Ll | Heterozygous | Long wings |
| ll | Homozygous l (Homozygous recessive) | Vestigial wings |

**Table 1.1**
The genotypes and phenotypes of wing length in *Drosophila melanogaster*

**Figure 1.13**
A cross between a homozygous
long-winged fruit fly and a
homozygous vestigial-winged
fruit fly to show the expected
phenotypes in the offspring (1)
and offspring (2) generations.
In a Unit test, you would not be
expected to draw the fruit flies

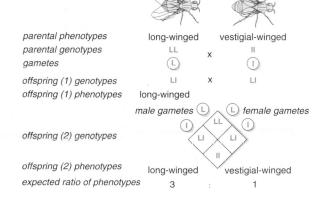

Figure 1.13 represents a cross between a homozygous long-winged fly
and a homozygous vestigial-winged fly. Flies in the offspring (1)
generation are all heterozygous and, since the L allele is dominant, are

**Figure 1.14**
Snapdragon plants (*Antirrhinum majus*) showing variation in flower colour. The gene for flower colour has two codominant alleles, one for red flowers and one for white flowers

long-winged. Figure 1.13 also shows that if two of the heterozygous offspring (1) generation are mated together, they produce an offspring (2) generation in which the expected ratio of long-winged to vestigial-winged flies is 3:1. This ratio is typical of a cross between heterozygotes in a monohybrid cross where one of the alleles is dominant. (Note that in crosses such as this, where the parents are homozygous, the offspring (1) and offspring (2) generations might also be referred to as the F1 and F2 generations in a Unit test.)

A checkerboard (called a **Punnett diamond**) has been used to represent all the possible genotypes that could result from the random fusion of gametes to form the offspring (2) generation in Figure 1.13. This notation helps to simplify the diagram: you should find it particularly helpful later in this chapter when looking at the possible genotypes that could result in the offspring (2) generation of a dihybrid cross.

**Q**   7   What name refers to an allele that always shows its effect in the phenotype?

## Monohybrid inheritance with codominance

In the example of monohybrid inheritance used above, one allele of the gene for wing length was dominant over the other allele of the gene for wing length. The alleles of some genes always show their effect in the phenotype: neither is dominant or recessive. When this occurs, the alleles are said to be **codominant**. Figure 1.14 shows flowering plants called snapdragons (*Antirrhinum majus*). Although the flowers on each plant are red, pink or white, flower colour in *Antirrhinum* is controlled by a single gene with two alleles. Figure 1.15 represents a cross between a red-flowered plant and a white-flowered plant. Just as was the case with *D. melanogaster* in Figure 1.13, the homozygous parents produce an offspring (1) generation that are all heterozygous. Because the red-flower and white-flower alleles are codominant, these heterozygotes have pink flowers. When two of these pink-flowered plants are crossed, an offspring (2) generation is produced that contains red, pink and white-flowered plants in the expected ratio of 1:2:1. This ratio is typical of a cross between two heterozygotes in a monohybrid cross with codominance.

**Figure 1.15**
A cross between a pure-breeding (homozygous) red-flowered snapdragon and a pure-breeding (homozygous) white-flowered snapdragon to show the expected phenotypes in the offspring (1) and offspring (2) generations

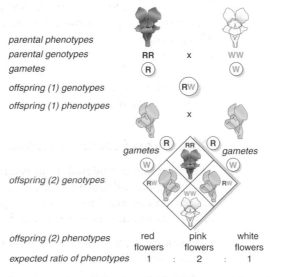

| | | | |
|---|---|---|---|
| parental phenotypes | | | |
| parental genotypes | **RR** | x | **WW** |
| gametes | (R) | | (W) |
| offspring (1) genotypes | | (RW) | |
| offspring (1) phenotypes | | x | |
| | gametes (R) **RR** (R) gametes | | |
| | (W) | | (W) |
| offspring (2) genotypes | (RW) | **WW** | (RW) |
| offspring (2) phenotypes | red flowers | pink flowers | white flowers |
| expected ratio of phenotypes | 1 : | 2 : | 1 |

**Q**   8   Explain why snapdragons with the genotype RW have pink flowers and not red flowers.

**Extension box 2**

## Test crosses

Sometimes, we need to know whether an organism showing the effects of the dominant allele of a particular gene is homozygous dominant or heterozygous. To find out, we perform a **test cross**. This involves crossing the organism of unknown genotype with a number of homozygous recessive individuals. The homozygous recessive individuals produce gametes that contain only recessive alleles. If the organism of unknown genotype is homozygous, all its gametes will contain the dominant allele and all the offspring will be heterozygous. We will recognise this because all the offspring of the test cross will show the effect of the dominant allele (Figure 1.16a). If the unknown organism is heterozygous, half of its gametes will contain the recessive allele and half of its offspring would be expected to be homozygous recessive and show the effects of the recessive allele (Figure 1.16b).

**Figure 1.16**
A test cross is used to find the genotype of an organism that shows the effects of a dominant allele. In this case, a fruit fly with long wings is crossed with a number of homozygous recessive vestigial-winged mates. The ratio of long-winged and vestigial-winged flies in the offspring show whether the long-winged fly was homozygous (a) or heterozygous (b)

**(a) if long-winged parent is homozygous**

| | | | |
|---|---|---|---|
| *parental phenotypes* | long-winged | x | vestigial-winged |
| *parental genotypes* | LL | | ll |
| *gametes* | (L) | | (l) |
| *offspring (1) genotypes* | | Ll | |
| *offspring (1) phenotypes* | | long-winged | |

**(b) if long-winged parent is heterozygous**

| | | | |
|---|---|---|---|
| *parental phenotypes* | long-winged | x | vestigial-winged |
| *parental genotypes* | Ll | | ll |
| *gametes* | (L) (l) | | (l) |
| *offspring (1) genotypes* | Ll | | ll |
| *offspring (1) phenotypes* | long-winged | | vestigial-winged |
| *expected ratio of phenotypes* | 1 | : | 1 |

A test cross is carried out to determine the genotype of organisms that are commercially important. We are unlikely to want to know whether a long-winged fruit fly is homozygous or heterozygous but we would use a test cross to determine whether a bull showing a desirable phenotype was homozygous, in which case its sperm could be used for artificial insemination, or heterozygous, which would make it less valuable for artificial insemination.

## Monohybrid cross with multiple alleles

The two genes considered so far each had two alleles. Many genes have more than two alleles. You learned about the human ABO blood groups in your AS course. These blood groups are controlled by an immunoglobulin gene (represented I) that has three alleles:

- $I^A$ results in the formation of antigen A on the plasma membrane of red blood cells

- $I^B$ results in the formation of antigen B on the plasma membrane of red blood cells

- $I^O$ results in the formation of neither antigen A nor antigen B on the plasma membrane of red blood cells.

Although there are three alleles of the **I** gene, no more than two can be present in a diploid cell. The possible diploid genotypes involving these three alleles and the phenotypes that result from each are shown in Table 1.2. From this, you can see that alleles $I^A$ and $I^B$ are codominant but that both are dominant over the recessive $I^O$ allele.

| Genotype | Antigens present on plasma membrane of red blood cells | Phenotype |
|---|---|---|
| $I^A I^A$ | Antigen A | Blood group A |
| $I^A I^B$ | Antigen A and antigen B | Blood group AB |
| $I^A I^O$ | Antigen A | Blood group A |
| $I^B I^B$ | Antigen B | Blood group B |
| $I^B I^O$ | Antigen B | Blood group B |
| $I^O I^O$ | Neither antigen A nor antigen B | Blood group O |

Table 1.2
The genotypes that result in the human ABO blood groups

**Q** 9 What is the evidence in Table 1.2 that:
(a) the $I^O$ allele is recessive
(b) the $I^A$ and $I^B$ alleles are codominant?

Figure 1.17
The human ABO blood group shows monohybrid inheritance with multiple alleles, i.e. more than two alleles of the controlling gene

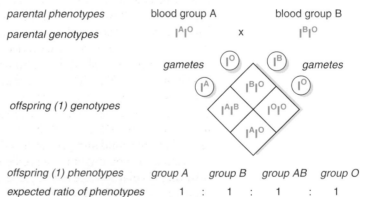

Figure 1.17 represents a cross between a heterozygous person of blood group A and a heterozygous person of blood group B. The interrelationships between the $I^A$, $I^B$ and $I^O$ alleles produce an interesting range of possible phenotypes amongst the offspring. Notice that two of them are unlike either parent.

## Dihybrid inheritance

So far, we have looked at the inheritance of one gene at a time. In **dihybrid inheritance**, we consider the inheritance of two genes that occur at two different loci. This might sound difficult, but we follow exactly the same format as we have used in representing monohybrid crosses.

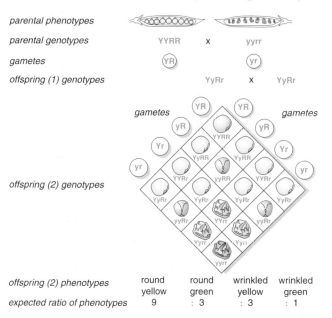

*parental phenotypes*

*parental genotypes*　　　YYRR　　x　　yyrr

*gametes*　　　　YR　　　　　yr

*offspring (1) genotypes*　　　YyRr　　x　　YyRr

*offspring (2) genotypes*

*offspring (2) phenotypes*　　round yellow　round green　wrinkled yellow　wrinkled green
*expected ratio of phenotypes*　　9　:　3　:　3　:　1

**Figure 1.18**
The inheritance of colour and shape of peas is an example of dihybrid inheritance. The colour and shape of seeds were just two of the characteristics of pea plants investigated by Gregor Mendel, the Austrian monk who discovered the principles of inheritance

Let's consider one of Mendel's crosses with pea plants. Peas are the seeds of pea plants. Mendel looked at the inheritance of two phenotypic features of peas: seed colour and seed shape. The gene for seed colour has two alleles: the allele for yellow colour (**Y**) is dominant over that for green colour (**y**). The gene for seed shape also has two alleles: the allele for round seeds (**R**) is dominant over the allele for wrinkled seeds (**r**). Figure 1.18 represents a cross between a pea plant that is homozygous dominant for both genes (i.e. has the genotype **YYRR**) and one that is homozygous recessive for both genes (i.e. has the genotype **yyrr**) and in which two of the offspring (1) plants were bred together. All the offspring (1) generation are heterozygous for the two genes (**YyRr**) and show the effects of the dominant alleles, i.e. seeds that are round and yellow. It is the offspring (2) generation that is interesting. As you might have predicted from Figure 1.8, meiosis in each offspring (1) plant will result in gametes with four different possible genotypes – **YR**, **yR**, **Yr** and **yr** – in equal numbers. The random fertilisation of these gametes results in the sixteen different combinations in the offspring (2) generation shown in Figure 1.18. Of these sixteen possible combinations, 9 are for round yellow peas, 3 are for round green peas, 3 are for wrinkled yellow peas and 1 is for wrinkled green peas. This ratio of 9:3:3:1 is characteristic of the offspring (2) or F2 generation in a dihybrid cross and will help you to recognise a dihybrid cross in a Unit test.

**Q** 10 Give the ratio of F2 phenotypes you would expect from a cross involving:
(a) monohybrid inheritance
(b) dihybrid inheritance.

**Synoptic
extension box**

## A molecular model of dominance

You learned in your AS course that:

- many proteins are single polypeptides

- a gene is a length of DNA that controls the production of a particular polypeptide by a cell

- enzymes are proteins

- metabolic pathways are controlled by enzymes.

We can put this information together to produce a molecular model of dominance and codominance.

Domestic rabbits eat plants that contain a yellow, fat-soluble pigment, called xanthophyll. This yellow pigment would accumulate in the body fat of rabbits, colouring their fat yellow. This does not normally happen because rabbits produce an enzyme that breaks down xanthophyll to a colourless compound. As a result, their body fat is white. The enzyme that breaks down xanthophyll is coded by the 'normal' allele of a gene that is present in the genotype of most rabbits.

'normal' allele $\Rightarrow$ active enzyme

xanthophyll $\rightarrow$ colourless compound

There is also a 'defective' allele of this gene that codes for an inactive enzyme.

'defective' allele $\Rightarrow$ inactive enzyme

xanthophyll $\rightarrow$ no reaction

You should remember from your AS course that enzymes are used time and again in cells. Provided a rabbit has the 'normal' allele for the xanthophyll enzyme, it produces enough enzyme to break down all the xanthophyll from the rabbit's diet. Rabbits that are heterozygous for this gene, as well as rabbits that are homozygous for the 'normal' allele, will have white fat. However, rabbits that are homozygous for the 'defective' allele will produce no active enzyme. As a result, xanthophyll is not broken down and dissolves in their body fat, colouring it yellow.

This leads to a general conclusion. A dominant allele on one homologous chromosome of a pair is capable of synthesising enough of the enzyme for which it codes to result in a phenotype that is indistinguishable from the phenotype that results if the dominant allele is present on both homologous chromosomes in a pair. A recessive allele codes for the production of an inactive enzyme or no enzyme at all, i.e. some recessive alleles are deletions.

The $I^A$ and $I^B$ alleles of the human ABO blood groups code for slightly different active enzymes. The A enzyme attaches galactosamine to glycolipid molecules in the plasma membranes of red blood cells: this forms the A antigens. The B enzyme attaches galactose (a monosaccharide) to glycolipid molecules in the plasma membranes of red blood cells: this forms the B antigens. If a person has the genotype $I^A I^B$, he will produce both active enzymes so that the plasma membranes of his red blood cells will have both A and B antigens. The $I^O$ allele codes for no active enzyme. As a result, neither antigen is formed.

## Sex linkage

We saw earlier in the chapter that human sex is inherited as a result of the two sex chromosomes, **X** and **Y**. We also saw that the **X** chromosome is longer than the **Y** chromosome and carries a large number of genes that are unrelated to sex and are absent from the **Y** chromosome. **Sex linkage** is the name given to the situation in which a phenotypic characteristic is inherited as a result of a gene on the **X** chromosome that is absent from the **Y** chromosome. Sex linkage is like dihybrid inheritance, since we are looking at the inheritance of sex at the same time as that of another phenotypic feature.

**Figure 1.19**
This pedigree shows the inheritance of haemophilia in humans. Note that haemophilia is more common in males but is never passed from father to son: these are characteristics of sex-linkage

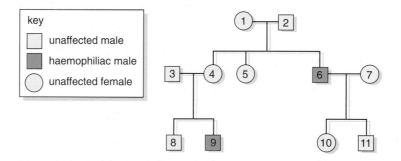

Figure 1.19 represents part of a pedigree of a family that is affected by **haemophilia**, a disease in which the clotting time of blood is much longer than usual. You need to become familiar with interpreting diagrams like this. The circles represent females and the squares represent males. Individuals 1 and 2, both unaffected by haemophilia, had three children, two unaffected daughters (individuals 4 and 5) and a son who suffered from haemophilia (individual 6). As an adult, individual 4 had two sons by an unaffected man (individual 3): one of these was unaffected but one (individual 9) suffered from haemophilia. As an adult, individual 6 had a daughter and a son by an unaffected woman (individual 7): neither child was affected by haemophilia.

The pattern of inheritance of haemophilia shown in Figure 1.19 is typical of sex-linked inheritance. A sex-linked characteristic is:

- more common in males than in females

- never passed from father to son.

**Q** 11  **Why is haemophilia never passed from a father to his son?**

Table 1.3 represents the sex chromosomes involved in the inheritance of haemophilia. On each **X** chromosome is a gene for blood clotting that is totally absent from the **Y** chromosome. The gene has two alleles, the dominant **H** that results in a rapid blood clotting time and the recessive **h** that results in a slowed clotting time. The symbols from Table 1.3 have been used in Table 1.4 to represent the possible genotypes and phenotypes involved in the inheritance of haemophilia.

| Symbol | Description |
| --- | --- |
| $X^H$ | An **X** chromosome that carries the dominant **H** allele of the blood clotting gene |
| $X^h$ | An **X** chromosome that carries the recessive **h** allele of the blood clotting gene |
| Y | A **Y** chromosome. This small chromosome does not carry a blood clotting gene |

Table 1.3
The symbols used to represent the inheritance of haemophilia in humans

| Genotype | Description of phenotype |
| --- | --- |
| $X^H Y$ | A male with rapid clotting time |
| $X^h Y$ | A haemophiliac male |
| $X^H X^H$ | A homozygous female with rapid clotting time |
| $X^H X^h$ | A heterozygous female 'carrier' who has rapid clotting time but carries the recessive allele |
| $X^h X^h$ | A homozygous haemophiliac female. (These are rare because they arise from a haemophiliac father and a mother who is a 'carrier'.) |

Table 1.4
The genotypes and phenotypes in human haemophilia. Note that the **Y** chromosome lacks the blood clotting gene. As a result, all males are described as hemizygous for this gene - whichever they inherit on the **X** chromosome from their mother is expressed in their phenotype

We can use this information to work out the genotypes of individuals in the family pedigree in Figure 1.19. Individual 6 is shown as a haemophiliac male: we know from Table 1.4 that he must have the genotype $X^h Y$. Males always inherit their **X** chromosomes from their mothers and their **Y** chromosomes from their fathers. Consequently, we know that individual 1 must have the $X^h$ chromosome in her genotype. Since she is not affected by haemophilia, she must have the genotype $X^H X^h$, in other words she is a carrier. Individual 2, an unaffected male, must have the genotype $X^H Y$.

**Q** 12 Explain why each of the following must have the genotype $X^HX^h$:
(a) individual 4
(b) individual 10.

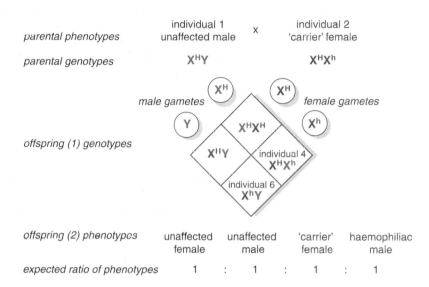

| parental phenotypes | individual 1 unaffected male | x | individual 2 'carrier' female |
|---|---|---|---|

parental genotypes $X^HY$ $X^HX^h$

male gametes $X^H$ $X^H$ female gametes

$Y$ $X^h$

offspring (1) genotypes $X^HX^H$

$X^HY$ individual 4 $X^HX^h$

individual 6 $X^hY$

| offspring (2) phenotypes | unaffected female | unaffected male | 'carrier' female | haemophiliac male |
|---|---|---|---|---|
| expected ratio of phenotypes | 1 : | 1 : | 1 : | 1 |

**Figure 1.20**
A cross showing the inheritance of haemophilia, using individuals 1, 2, 4 and 6 from Figure 1.19

Figure 1.20 shows how we can use our standard method for representing the cross between individuals 1 and 2. The genotype we have deduced for individual 4 ($X^HX^h$) and for individual 6 ($X^hY$) are shown in Figure 1.20. From the information given, we cannot deduce the genotype of individual 5.

## When offspring from experimental crosses do not exactly match expected ratios: the chi-squared ($\chi^2$) test

If you were to toss a perfectly balanced coin, there is a 50:50 chance that it will fall as 'heads'. If it fell 'tails', you would not be surprised - this could happen as a result of chance. If you tossed the coin in the air twice and on both occasions it landed 'tails', you would not be surprised - again, this could happen as a result of chance. If you tossed the coin in the air 200 times and it landed 'tails' every time, you would not accept that this was due to chance and would suspect that the coin was unevenly weighted or double-sided 'tails'. This illustrates two ideas. Firstly, we do not let chance variations from what we expect prevent us from accepting a hypothesis (in this case, that the coin is perfectly balanced and has a 'heads' side and a 'tails' side). Secondly, there is a point at which variations from what we expect are so great that we begin to doubt our hypothesis and look for another (that the coin is unevenly weighted or that both sides are the same).

Let's relate this idea to some of the crosses we have examined in this chapter. In Figure 1.11, we saw that there is a theoretical 1 in 2 chance of a human child being a boy or a girl. The chance factor here is that an egg cell with an **X** chromosome could be fertilised at random by either an **X**-carrying sperm or a **Y**-carrying sperm. You probably know families with two, three of four children who are all the same sex. This does not make us doubt our hypothesis that there is a 1 in 2 chance of each child being a boy. In fact, for every family with n children there will be $2^n$ possible permutations of births. For example, for every family of two children there will be four ($= 2^2$) permutations of births: boy-boy; boy-girl; girl-boy; girl-girl. On this basis, we would expect that one quarter of couples with two children would have two daughters, one quarter would have two sons and half would have a daughter and a son.

Humans have so few children that we never begin to doubt that chance is playing a role in their inheritance. This is not so with fruit flies, where a single mating can produce over two hundred offspring. Look back at the inheritance of wing length in *Drosophila melanogaster* in Figure 1.13. If 200 offspring were produced, we would expect 150 to have long wings and 50 to have vestigial wings. If there were actually 145 long-winged and 55 vestigial-winged, we would probably accept this was a chance variation from what we expected. If there were 50 long-winged and 150 vestigial-winged offspring, we would not accept this as a chance deviation from what we expected and would begin to doubt that our explanation of inheritance was true. But where do we draw the line? Two conventions help us here.

- We need a way of measuring the probability that what we observe is a chance deviation from what we expect. This is what the chi-squared test ($\chi^2$ test) does.

- We need a commonly agreed limit on what we accept is due to chance. For our purposes, we reject a hypothesis if the probability of our observations being a chance variation of what we expected to happen is less than 1 in 20 (5% or, as a decimal fraction, once its probability is less than 0.05). We say that the probability value of 0.05 is our **level of significance**.

To calculate $\chi^2$, we use the following formula, where **O** represents the observed value and **E** represents the expected value:

$$\chi^2 = \sum \frac{(O-E)^2}{E}$$

The Greek letter $\Sigma$ (pronounced 'sigma') means 'sum of' and shows that we must add up all the different calculated values of:

$$\frac{(O-E)^2}{E}$$

Suppose 200 offspring were produced in the cross shown in Figure 1.13. Anticipating a 3:1 ratio, we would expect 150 of them to be long-winged and 50 to be vestigial-winged. In fact, suppose we found that only 145 were long-winged and 55 were vestigial-winged. Is this a chance variation of what we expected? Table 1.5 shows how we would perform a $\chi^2$ test of our null hypothesis, i.e. that there is no difference between the ratio we observed and the 3:1 ratio that we expected.

| Component | Long-winged offspring | Vestigial-winged offspring |
|---|---|---|
| Expected (E) | 150 | 50 |
| Observed (O) | 145 | 55 |
| O – E | –5 | 5 |
| $(O-E)^2$ | 25 | 25 |
| $\dfrac{(O-E)^2}{E}$ | 0.17 | 0.5 |
| $\chi^2 = \sum \dfrac{(O-E)^2}{E}$ | 0.67 | |

**Table 1.5**
Calculation of $\chi^2$

From Table 1.5, the calculated value of $\chi^2$ is 0.67. We now need to use Table 1.6 to see whether this value of $\chi^2$ represents a chance variation of what we expected or not. As there are 2 different categories of flies in our example ($n = 2$), there is $n - 1 = 1$ degree of freedom. The table shows that, with one degree of freedom, the probability of our observed ratio of long-winged:vestigial-winged flies occurring by chance is between 0.50 and 0.30. This is above our level of significance of 0.05, so we can accept the ratio as a chance variation of what we expected.

| Categories ($n$) | Degrees of freedom ($n-1$) | $\chi^2$ | | | | | | | |
|---|---|---|---|---|---|---|---|---|---|
| 2 | 1 | 0.016 | 0.15 | 0.46 | 1.07 | 2.71 | 3.84 | 5.41 | 6.64 |
| 3 | 2 | 0.21 | 0.71 | 1.39 | 2.41 | 4.61 | 5.99 | 7.82 | 9.21 |
| 4 | 3 | 0.58 | 1.42 | 2.37 | 3.67 | 6.25 | 7.82 | 9.84 | 11.34 |
| Probability that chance alone could produce the deviation from what was expected | | 0.90 | 0.70 | 0.50 | 0.30 | 0.10 | 0.05 | 0.02 | 0.01 |

**Table 1.6**
Table of $\chi^2$ values. To accept our null hypothesis, that there is no significance between our expected and observed values, we need to find a value of $\chi^2$ that is associated with a probability of 0.05 or more. If the probability is less than 0.05, we reject our null hypothesis

# Summary

- During meiosis, cells with two copies of each chromosome (diploid cells) divide to form cells with only one copy of each chromosome (haploid cells).

- The separation of homologous chromosomes takes place during the first meiotic division. During the second meiotic division, sister chromatids of each chromosome are separated into daughter cells.

- Because of the way that the members of different pairs of homologous chromosomes separate independently of each other during the first meiotic division (independent assortment of homologous chromosomes), meiosis produces daughter cells that are genetically different from each other. In general, if there are $n$ pairs of homologous chromosomes, independent assortment will produce $2^n$ different combinations of chromosomes in the daughter cells. This genetic variation is increased because chromatids break at points where they intertwine (chiasma) and join to the other homologue.

- Monohybrid crosses involve the inheritance of a character controlled by a single gene. Dihybrid crosses involve the inheritance of characters controlled by two genes that occur at two different loci.

- Any diploid cell has two copies of the gene controlling a single character. If both versions of the gene (called alleles) are the same, this individual has a homozygous genotype. If the two alleles of the gene are different, the individual has a heterozygous genotype.

- In representing genetic crosses, we use a standard format that shows: phenotype of parents, genotype of parents, gametes, genotype of offspring (1) generation, expected frequency of phenotype of offspring (1) generation. If the parents are pure-breeding (homozygous), the offspring (1) generation is sometimes called the F1 generation and the offspring (2) generation is sometimes called the F2 generation.

- An allele of a gene is said to be recessive if it fails to exert its effect in a heterozygote. The allele of the same gene that exerts its effect in the heterozygote is said to be dominant. The alleles are codominant if they both show their effect in a heterozygote.

- Sex is determined by sex chromosomes. A character unrelated to sex is sex-linked if it is controlled by a gene whose locus is on the female sex chromosome.

- The chi-squared test ($\chi^2$ test) helps us to determine whether observed results fit the expectations we had based on a null hypothesis. As a general rule, we accept any variation from what we expected so long as its probability is $\geqslant 0.05$.

# Assignment

## Ladybird, ladybird

Ladybirds, such as those shown in Figure 1.21, are beetles. They are common insects and their distinctive colour patterns make them quite easy to identify. Some species are also very variable in appearance. Many of these variations are genetically controlled. In this assignment we will look at the inheritance of some of the colour patterns found on the wing-cases of ladybirds. Before you start work, look at the examples of crosses shown in this chapter and make sure you know how to use a genetic diagram to explain the results of a particular cross. The crosses at the start of this assignment are very straightforward but you may find some of those towards the end much more challenging!

The 2-spot ladybird is a much more variable species than the 7-spot ladybird shown in Figure 1.21. Figure 1.22 shows two varieties of this ladybird. These varieties are controlled by a single gene with two alleles. The allele for black wing-cases, **B**, is dominant to that for red wing-cases, **b**.

1  Use a genetic diagram to show the offspring you would expect from a cross between a red ladybird and a ladybird heterozygous for these alleles.

*(3 marks)*

Three more varieties of this ladybird are shown in Figure 1.23. These varieties are controlled by another gene. It has two alleles. The typica allele ($W^T$) and the annulata allele ($W^R$) are codominant. The heterozygote is intermediate in appearance.

2  (a)  Construct a table to show the possible genotypes of each of the three ladybirds shown in Figure 1.23.

*(2 marks)*

   (b)  Predict the ratio of phenotypes of the offspring from a cross between two intermediate ladybirds. Use a genetic diagram to explain your answer.

*(2 marks)*

The 2-spot ladybird also occurs in the United States. We will look now at some work carried out on American 2-spot ladybirds nearly a hundred years ago. Figure 1.24 shows five varieties of this ladybird. These varieties were found to be controlled by a single gene with five alleles: $W^B$ (*bipunctata*), $W^M$ (*melanopleura*), $W^A$ (*annectans*), $W^C$ (*coloradensis*) and $W^H$ (*humeralis*).

**Figure 1.21**
Ladybirds are common insects but they are not often seen in large numbers. However, swarms can contain as many as 400 insects in a square metre. Those pictured here are 7-spot ladybirds

Black wing cases    Red wing cases

**Figure 1.22**

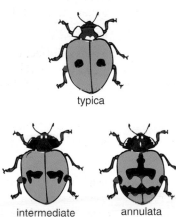

typica

intermediate    annulata

**Figure 1.23**

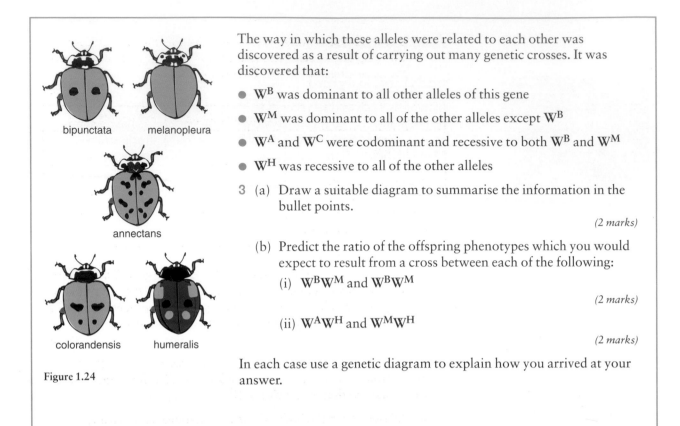

bipunctata

melanopleura

annectans

colorandensis

humeralis

Figure 1.24

The way in which these alleles were related to each other was discovered as a result of carrying out many genetic crosses. It was discovered that:

- $W^B$ was dominant to all other alleles of this gene
- $W^M$ was dominant to all of the other alleles except $W^B$
- $W^A$ and $W^C$ were codominant and recessive to both $W^B$ and $W^M$
- $W^H$ was recessive to all of the other alleles

3  (a)  Draw a suitable diagram to summarise the information in the bullet points.

*(2 marks)*

(b)  Predict the ratio of the offspring phenotypes which you would expect to result from a cross between each of the following:

(i)  $W^B W^M$ and $W^B W^M$

*(2 marks)*

(ii)  $W^A W^H$ and $W^M W^H$

*(2 marks)*

In each case use a genetic diagram to explain how you arrived at your answer.

# Examination questions

1  A queen honey bee can lay both fertilised and unfertilised eggs. Fertilised eggs develop into diploid females and unfertilised eggs develop into haploid males. The diagram shows the formation of gametes in female bees and in male bees.

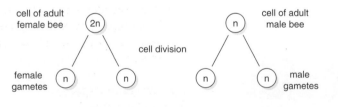

cell of adult female bee — 2n

cell of adult male bee — n

cell division

female gametes — n    n

n    n — male gametes

(a)  Giving a reason for your answer in each case, name the type of cell division in the bee that produces:

(i)  female gametes

(ii)  male gametes

*(1 mark)*

(b) The table shows some features which contribute to variation in the offspring of bees. Complete the table with a tick if the feature may contribute or a cross if it does not.

*(2 marks)*

| Feature | Female offspring | Male offspring |
|---|---|---|
| Crossing over | | |
| Independent segregation of chromosomes | | |
| Random fusion of gametes | | |

(c) Body colour in bees is determined by a single gene. The allele **B** for yellow body is dominant to the allele **b** for black body. Explain why, in the offspring of a mating between a pure-breeding black female and a yellow male, all the males will be black.

*(1 mark)*

2  In cats, one of the genes for coat colour is present only on the X chromosome. This gene has two alleles. The allele for ginger fur, $X^B$, is dominant to that for black fur, $X^b$.

(a) All the cells in the body of a female mammal carry two X chromosomes. During an early stage of development, one of these becomes inactive and is not expressed. Therefore female mammals have patches of cells with one X chromosome expressed and patches of cells with the other chromosome expressed. Tortoiseshell cats have coats with patches of ginger and patches of black fur.

   (i)  What is the genotype of a tortoiseshell cat?

   *(1 mark)*

   (ii) Explain why there are no male tortoiseshell cats.

   *(1 mark)*

(b) A cat breeder who wished to produce tortoiseshell cats crossed a black female cat with a ginger male. Complete the genetic diagram and predict the percentage of tortoiseshell kittens expected from this cross.

Parental phenotypes:      black female     ginger male
Parental genotypes:
Gametes:
Offspring genotypes:
Percentage of tortoiseshell kittens:

*(3 marks)*

3  Night blindness is a condition in which affected people have difficulty seeing in dim light. The allele for night blindness, **N**, is dominant to the allele for normal vision, **n**. (These alleles are *not* on the sex chromosomes.)

The diagram shows part of a family tree showing the inheritance of night blindness.

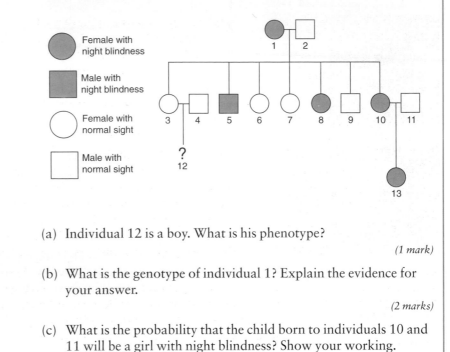

(a)  Individual 12 is a boy. What is his phenotype?

*(1 mark)*

(b)  What is the genotype of individual 1? Explain the evidence for your answer.

*(2 marks)*

(c)  What is the probability that the child born to individuals 10 and 11 will be a girl with night blindness? Show your working.

*(2 marks)*

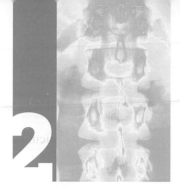

# Variation and Selection

**Figure 2.1**
These male emperor penguins spend the coldest two months of the Antarctic winter incubating their eggs. Each male has a single egg balanced on his feet and covered by the folds of feathered skin from his abdomen. He will stay like this for about two months, during which time he will lose up to one-third of his body mass

Did you spend part of today standing in a cold, wet bus queue wishing you had stayed at home? If so, spare a thought for the poor penguins in Figure 2.1. These penguins are male emperor penguins (*Aptenodytes forsteri*). They are incubating their eggs. Each male balances a single egg on his feet and is keeping it warm by covering it with folds of feathered skin from his abdomen. Despite the fact that the temperature can fall to −40 °C and wind speeds can exceed 160 km per hour, these males will stay huddled together, as you see them in Figure 2.1, for two months.

Look carefully at the birds in Figure 2.1. Can you distinguish one from another? The chances are that you cannot; one emperor penguin looks much the same as any other. When the females return after spending two months feeding at sea, they will return to their own mates. In fact, emperor penguins spend most of the year at sea. When they return to the ice around Antarctica to breed, they seek out their mates from the previous year. Obviously, emperor penguins do recognise each other.

Since emperor penguins reproduce sexually, there is genetic variation between members of the species. In sexually reproducing populations, only identical twins are genetically identical to each other. A female only ever lays one egg in a breeding season, so there are no pairs of identical twins amongst emperor penguins. It is this genetic variation that enables emperor penguins to recognise each other. We are good at recognising genetic differences between humans, so that no two humans look identical to us. Our inability to recognise genetic differences between emperor penguins does not mean that they are not there.

In this chapter, you will learn about some of the causes of variation and appreciate why genetic variation is so common in sexually reproducing populations. You will also see how variation can lead to genetic changes in populations, through natural selection, and begin to use simple statistical techniques to investigate variation.

## Investigating variation

In this part of the chapter, you will learn about ways in which we can collect, represent and analyse quantitative data. You might use some of these techniques during your coursework investigations.

### Frequency distributions

In Chapter 1, we looked at the way in which some phenotypic characteristics are inherited. Look back to Figure 1.13, which shows the inheritance of wing length in *Drosophila melanogaster*. If a pair of *Drosophila* reproduced exactly as you would expect from Figure 1.13, there would be three times as many long-winged flies in the offspring (2)

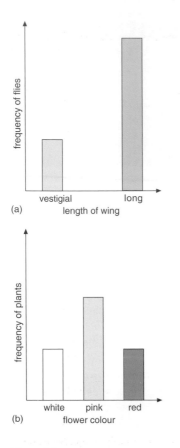

(a)

(b)

**Figure 2.2**
(a) A frequency distribution
of the expected wing lengths of
the offspring (2) generation of
flies produced by the cross in
Figure 1.13
(b) A frequency distribution of
the expected flower colours in the
offspring (2) generation of
snapdragons produced by the
cross in Figure 1.15

generation as there were vestigial-winged flies. We can plot these expected results on a bar graph. This has been done in Figure 2.2(a). On the horizontal axis, we have the characteristic being measured, wing length, and on the vertical axis we have the frequency of organisms in each category of wing length. Because this type of graph shows the frequency of organisms within each of several measured categories, it is called a **frequency distribution**. Because we have only two wing-length categories – vestigial and long – there are only two bars on our frequency distribution. Now look back to Figure 1.15, which shows the inheritance of flower colour in snapdragons (*Antirrhinum majus*). If a pair of snapdragon plants reproduced exactly as you would expect from Figure 1.15, there would be equal numbers of plants with red flowers and white flowers but twice as many plants with pink flowers. Figure 2.2(b) shows a frequency distribution of these expected results. Instead of the two categories of wing length that we had in Figure 2.2(a) we now have three categories of flower colour and the frequency distribution has a different shape.

It is easy to judge whether the wing length of a fly is vestigial or long or to judge whether a flower is red, pink or white. This is because these characteristics are quite distinct from each other. Data like these are described as **discrete data**. A frequency distribution of discrete data always shows **discontinuous variation**. However, data are not always discrete. Imagine a student wished to carry out an investigation of the height of students in two of her study groups. Although some people are tall and others are short, there is not a single 'tall' category or a single 'short' category. The person conducting the investigation would have to decide the size categories (called class intervals) she would use. In doing so, she might choose the same class intervals as you or different class intervals, it would not much matter.

Table 2.1 shows how this student might record the heights of her fellow students. The left-hand column shows the class intervals she has chosen, in this case she has chosen 2 cm intervals. In the middle column, she recorded her raw data. She has used a common convention for doing this. Each time she measured a height within a particular class interval, she wrote 'I' in the appropriate row in the middle column. However, every fifth time she drew a diagonal line through the previous four 'Is'. This divides her raw data into groups of five, making counting easier. Finally, in the right-hand column, she totalled the number of students in each class interval. Figure 2.3 shows the frequency distribution plotted from the data in Table 2.1. Once again, the bars represent the number of counts in each class interval. However, a curve has been added to the frequency distribution to show the general trend. This trend shows that measurements of human height are not discrete data; instead they are continuous data. A frequency distribution of continuous data, like the curve in Figure 2.3, shows **continuous variation**.

| Class interval/mm | Tally count of number of students | Number of students |
|---|---|---|
| 1480–1499 | I | 1 |
| 1500–1519 | III | 3 |
| 1520–1539 | IIII | 4 |
| 1540–1559 | IНΤ I | 6 |
| 1560–1579 | IНΤ II | 7 |
| 1580–1599 | IНΤ IНΤ | 10 |
| 1600–1619 | IНΤ III | 8 |
| 1620–1639 | IIII | 4 |
| 1640–1659 | III | 3 |
| 1660–1679 | II | 2 |
| 1680–1699 | I | 1 |

**Table 2.1**
A method for recording raw data about the height of students in study groups. Before the investigation, a decision was made about the class interval. A tally was recorded in the middle column, grouping counts into fives for ease of counting. The final column was completed at the end of the investigation

**Figure 2.3**
A frequency distribution plotted from the data in Table 2.1. The curve shows the general trend of the data

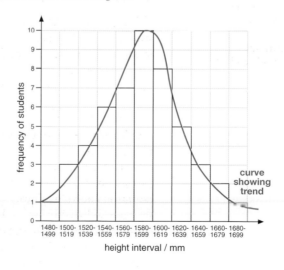

## The effect of chance and bias on data collection

In the examples used above, we looked at samples of a population – the offspring (2) generation from a single cross or the people who happen to be in the same study groups. These samples are sub-groups of the total population and might not be truly representative of that population. Even measuring the height of each person in her study groups was a sampling process. Had our student made the same measurement several times, she would probably have found that she recorded different heights for the same people.

Variation in sampling can come about in three ways:

● the differences between samples reflect real differences in the populations from which they came

● chance – what we would commonly call luck

● sampling bias – the investigator, knowingly or unknowingly chooses samples.

When we investigate populations, we want to be sure that any differences we find reflect real differences in the populations we are sampling.

We eliminate chance by taking several samples. A single, small sample is unlikely to be representative of the population from which it came. This is why you are encouraged to use the average of several readings during practical work or to combine results from the group and take the average values. The number of samples we take is limited by the time we have available. In laboratory work, you are probably encouraged to take three samples by setting up three replicates of an experiment or by taking three readings. Biologists often use more than three samples when they want to have confidence in their data.

We eliminate bias by removing human choice from the sampling method. Methods that do this ensure **random sampling**. You are expected to carry out fieldwork, during which you collect quantitative data, at least once in your A2 course. You will use random sampling methods to collect this data. These might involve drawing a grid over a map of the sample area, numbering squares and then choosing the numbers of the squares to be sampled from a table of random numbers. It will not involve throwing a square quadrat frame over your shoulder and sampling where it falls. You can read more about sampling methods in Chapter 4.

**Q** 1 Why would throwing a quadrat frame over your shoulder *not* be a suitable sampling method during fieldwork?

## The normal distribution

The normal distribution is a special type of frequency distribution. Figure 2.4 shows a graph of a normal distribution, called a normal distribution curve. Like any frequency distribution, it has the class characteristic being investigated on the horizontal axis and the number of individuals on the vertical axis. The curve has a symmetrical bell shape. It also has important mathematical properties, which are:

● its most frequent value (mode); middle value (median) and average value (mean) are the same

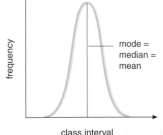

mode = median = mean

**Figure 2.4**
A normal distribution curve. This is a special type of frequency distribution in which the mode, median and mean are identical

- it is symmetrical – 50% of the values are above the mean and 50% of the values are below the mean

- 95% of its values are within two standard deviations of the mean.

Let's look at these concepts in a little more detail. Table 2.2 shows measurements that were made on the length of twenty-seven plant leaves. The lengths have been arranged in rank order. The **mode** is the measurement that occurs most often. If plotted on a frequency distribution, it will be the highest point in the bar chart or graph. In Table 2.2, the most common leaf length is 3.8 cm – this is the mode of these data. The **median** is the middle number in the ranked list. In Table 2.2, we have twenty-seven measurements. The middle measurement, with thirteen above it and thirteen below it, is 3.8 cm. In this case, as well as being the modal value, 3.8 cm is also the median.

The arithmetic mean is written as $\overline{x}$ (pronounced 'x bar'). It is calculated as

$$\text{Mean} = \frac{\text{sum of all the measurements}}{\text{number of measurements}}$$

In mathematical notation this is written as

$$\overline{x} = \frac{\sum x}{n}$$

We can use this formula to calculate the arithmetic mean of the measurements in Table 2.2.

Sum of all the measurements = 100.7

Number of measurements = 27

Mean length = 100.7 / 27 – 3.7

**Q** 2 Do the data in Table 2.2 follow a normal distribution? Explain your answer.

| Length of leaf/cm | | | | | | | | |
|---|---|---|---|---|---|---|---|---|
| 2.9 | 3.0 | 3.1 | 3.1 | 3.3 | 3.4 | 3.5 | 3.6 | 3.6 |
| 3.7 | 3.7 | 3.7 | 3.8 | 3.8 | 3.8 | 3.8 | 3.8 | 3.9 |
| 3.9 | 3.9 | 4.0 | 4.0 | 4.1 | 4.2 | 4.3 | 4.3 | 4.5 |

Table 2.2
The length of the leaves in a sample of plant leaves

## Extension box 1

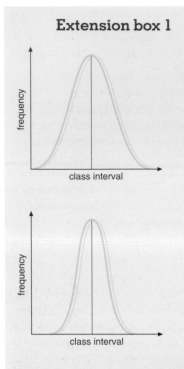

**Figure 2.5**
Although these are both normal distribution curves, one has a greater spread than the other. The curve with the greater spread has a greater standard deviation than the thinner curve

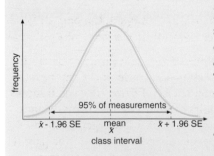

**Figure 2.6**
A normal distribution curve showing the mean ($\bar{x}$) and 1.96 standard errors (SE) either side of the mean. 95% of the measurements lie within the range mean ± 1.96 SE

## Standard deviation

Although you will not be expected to calculate standard deviations and standard errors in a Unit test; you are expected to be familiar with their meaning and their use. You might also wish to calculate these measures using data that you collect in your coursework.

Figure 2.5 shows two normal distribution curves. Although they have the same mean, median and mode, they have a different spread of measurements. We represent the spread of measurements about the mean of a normal distribution curve by a value called the **standard deviation**. We calculate the standard deviation in the following way:

1. square each measurement ($x^2$)
2. add all the squared values together ($\sum x^2$)
3. divide this total by the number of measurements made $\dfrac{\sum x^2}{n}$
4. square the mean of the measurements ($\bar{x}^2$)
5. subtract the result of step 4 from the result of step 3 $\left( \dfrac{\sum x^2}{n} - \bar{x}^2 \right)$
6. take the square root of step 5. This is the standard deviation. Written as mathematical equation, the above steps becomes:

$$\text{standard deviation, } s = \sqrt{\left( \dfrac{\sum x^2}{n} - \bar{x}^2 \right)}$$

When given together, the mean and standard deviation of a series of measurements is useful to us. It tells us the central value of the distribution and how spread out the individual measurements are from the central value. Whatever the value of the mean and standard deviation, 95% of the data lie within two standard deviations above or below the mean.

Because we calculate means and standard deviations from small samples, they are only estimates of the means and standard deviations of whole populations. We take account of this using another measure, called the standard error of the mean (or **standard error**, for short). The standard error is calculated using the standard deviation (s) and the number of measurements (*n*):

$$\text{Standard error (SE)} = \dfrac{s}{\sqrt{n}}$$

The standard error is important because it is used to define confidence limits. The true mean of a whole population falls within the range ± 1.96 SE of the sample mean (Figure 2.6). You would be 95% confident that the mean of the population fell within the range ± 1.96 SE. If you look back to Table 1.6, you will see that we have already used this idea. Without knowing it then, the reason that we used the 5% (0.05) probability level in Table 1.6 was because we were rejecting any values that fell outside 1.96 SE of the mean of our sample.

**Q   3** Does a normal distribution curve that is thin have a greater or smaller standard deviation than one that is broad?

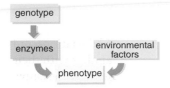

**Figure 2.7**
An organism's phenotype results from an interaction of genetic and environmental factors

**Figure 2.8**
The fur colour of this Himalayan rabbit shows how genetic and environmental factors interact to form the phenotype. The gene for black fur is active only in the colder parts of the rabbit's body

## The causes of variation

Variation exists between the members of any population. Variation is greater in populations of species that reproduce sexually than in those that reproduce asexually, but it is always there.

The differences we see between individuals in a population are differences in phenotype. In humans, characteristics such as height, eye colour, ABO blood group and presence or absence of cystic fibrosis are aspects of our phenotype. In other words, the **phenotype** describes the characteristics of an organism that we can see or detect. We learned about a number of phenotypes in Chapter 1, where we looked at inheritance. In Chapter 1, we used diagrams to show how alleles of genes were inherited to form genotypes, which then influenced phenotypes. However, the phenotype does not depend solely on an organism's genotype. As Figure 2.7 shows, the phenotype results from an interaction between genetic and environmental factors.

Look at the Himalayan rabbit in Figure 2.8, which illustrates this interaction. This rabbit has white fur over most of its body but has black fur on its nose, ears, tail and feet. Crossing two pure-breeding Himalayan rabbits will always result in offspring with these characteristics. Clearly, Himalayan rabbits must have a gene that codes for black fur, but why does it only show up in the nose, ears, tail and feet? The answer is simple. The gene for black fur codes for an enzyme that is inactivated at temperatures above 34 °C. Over most of the body, temperatures are above 34 °C so that the fur is white (albino). Only the nose, ears, tail and feet are cold enough for the enzyme to work.

**Q** 4 **Will cells that produce white hairs in the fur of Himalayan rabbits possess the gene for black colour?**

## Meiosis as a source of genetic variation

You learned about the process of meiosis in Chapter 1. On pages 11–13, you saw how meiosis generates genetic variation in several ways.

- **Independent assortment** of the alleles of genes on different pairs of homologous chromosomes results in genetic differences amongst the gametes produced by meiosis. Look back to Figure 1.6, which shows independent assortment of two pairs of homologous chromosomes. Figure 2.9 shows a similar diagram in which a single gene with two different alleles had been included on each chromosome. The parent cell, with a genotype of **AaBb** produces four genetically different daughter cells with genotypes **AB**, **Ab**, **aB** and **aB**.

- **Random fertilisation** of gametes from a single pair of parents results in genetic differences amongst the zygotes. Use Figure 1.18 and Figure 2.9 to help you work out the possible genotypes of offspring produced by two parents of genotype **AaBb**.

- **Crossing over** between members of a single pair of homologous chromosomes changes the combination of the alleles of genes found on those chromosomes. Look back to Figure 1.9 which shows chiasma

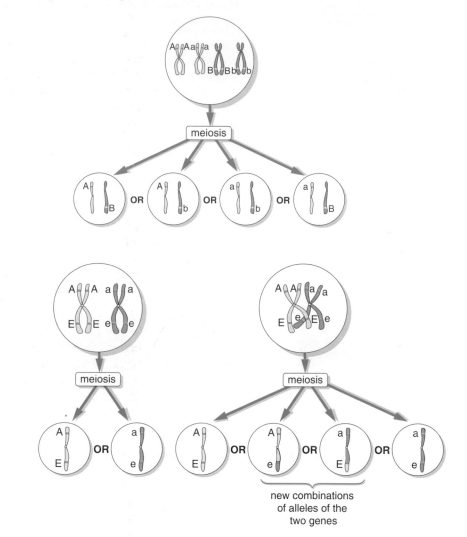

formation during meiosis. Figure 2.10 shows the effect of crossing over between the chromosomes from Figure 2.9. Genetic combinations are formed amongst the daughter cells that were not present in Figure 2.9.

**Figure 2.9**
During meiosis, an independent assortment of pairs of homologous chromosomes produces genetic differences amongst the daughter cells. In this diagram, chromosomes are represented as single lines carrying two genes, each of which has two alleles. Four different daughter cells are produced. Crossing over between pairs of homologous chromosomes can increase genetic variation

**Figure 2.10**
This diagram shows the effect of crossing over on the genetic combination of the chromosomes from Figure 2.9. Notice that new combinations (recombinants) are formed

**Q** 5 What genotypes would you expect among the offspring of a cross between two parents of genotype AaBb if the two genes are located on different chromosomes and no crossing over occurs?

## Gene mutation

During your AS course, you learned that:

● a gene is a length of DNA that carries the genetic code for the production of a particular polypeptide

● DNA is copied during the cell cycle, so that all cells produced by mitosis contain the same DNA as their parent cells

- when transcribed to a messenger RNA molecule, the genetic code is a sequence of bases (adenine, cytosine, guanine and uracil) that determines the sequence of amino acids in a polypeptide

- a combination of three mRNA bases (a codon) codes for a specific amino acid.

From this, you will realise that any change in a DNA base sequence can affect the polypeptide for which it carries the code. Changes in the DNA base sequence of genes are called **gene mutations**. They occur naturally during DNA replication at a rate of about 1 in $10^6$ bases copied. Some environmental factors, such as radiation, increase the gene mutation rate.

Base deletion and base substitution are two types of gene mutation.

- In a **base deletion**, at least one base is not copied during DNA replication. As a result, the new DNA lacks the deleted base or bases.

- In a **base substitution**, at least one base is copied wrongly during DNA replication. Remember that the DNA bases pair adenine with thymine and cytosine with guanine. These base pairs help to ensure the correct base sequence is maintained during DNA replication. During substitution, something interferes with this normal base pairing, so that the 'wrong' base becomes incorporated in a developing DNA strand.

Before looking at mRNA codons and the amino acids for which they code, let us use three-letter words to represent codons and sentences to represent genes. In Table 2.3, we have a sentence of three-letter words that makes sense to us. This represents a normal gene. Look at the third line in Table 2.3. A deletion has occurred, so that the combinations of three letters after the deletion no longer carry the intended meaning. The fourth line in Table 2.3 represents a substitution. One of the letters has been changed, so that the sentence no longer carries the intended meaning.

| Type of error | Combination of three-letter 'words' |
|---|---|
| None (normal message) | The cat saw the dog and ran off |
| Deletion (one letter lost) | The cat sat hed oga ndr ano ff |
| Substitution (wrong letter included) | The cat saw the log and ran off |

**Table 2.3**
This table represents two types of gene mutation. The three-letter 'words' represent base triplets and the 'sentences' represent genes. In a deletion mutation, the loss of one letter changes the meaning of the triplets that follow it. In a substitution mutation, only one of the base triplets is affected. The letters in red show affected triplets.

Just as the 'sentences' were changed in Table 2.3, base deletions and base substitutions change the genetic code. Table 2.4 shows some of the mRNA codons and the amino acids for which they code. The mRNA base triplet GGG codes for the amino acid glycine. A single change in the triplet can have no effect (e.g. GGA still codes for glycine) or can result in the incorporation of completely different amino acids in a polypeptide chain. Sickle-cell anaemia (which is described later in this chapter) and cystic fibrosis result from differences in protein structure of only one amino acid. Clearly, deletion and substitution can have profound effects on phenotypes.

| mRNA codon | Amino acid coded |
|---|---|
| GGG | glycine |
| GGA | glycine |
| GAG | glutamic acid |
| GCG | alanine |
| CGG | arginine |

Table 2.4
A few of the 64 possible mRNA codons and the amino acids for which they carry the code. A, C and G represent the organic bases adenine, cytosine and guanine, respectively

## Polygenic inheritance

In Chapter 1, we learned about the inheritance of discrete characteristics, for example wing length in *Drosophila melanogaster* and human ABO blood groups. As you saw, these characteristics are controlled by a single gene, with one or more alleles.

Many characteristics are not controlled by a single gene but by many genes, often located on different chromosomes. This is called **polygenic inheritance**. The different genes each contribute to the final phenotype. As a result, phenotypic characteristics controlled by polygenes have a different pattern of distribution from those controlled by a single gene. We saw these earlier in the chapter. Figure 2.2(a) showed the frequency distribution of a characteristic controlled by a single gene. An individual in the population of *Drosophila* either shows the dominant characteristic or the recessive characteristic. As a result, the frequency distribution has only two bars, representing discontinuous variation. Figure 2.4 showed a normal frequency distribution. This is typical of a characteristic controlled by many genes and represents continuous variation. Polygenic inheritance is involved in characteristics such as human height.

## Population genetics and allele frequencies

In Chapter 1, we considered the inheritance of genes from a single pair of parents to their offspring. In population genetics, we consider the genes in an entire population of organisms. Look back at Figure 1.13 which represents a monohybrid cross involving wing length in *Drosophila*. We used the symbol **L** to represent the dominant, long-wings allele of the gene for wing length and the symbol **l** to represent the recessive, vestigial-wings allele of the gene for wing length. In the offspring (2) generation the flies had one of three genotypes, **LL**, **Ll** or **ll**. Since these are the only possible genotypes for wing length, all the *Drosophila* in a population must have one of these three genotype. To represent the genotypes of all the individuals in a population of, say, 1000 *Drosophila* we would need to use 2000 symbols. These symbols would represent all the genes for wing length in that population. This is called the gene pool. In general, we can define a **gene pool** as the total number of alleles of a particular gene that are present in a population at a particular time.

If all the *Drosophila* in a population of 1000 individuals had the genotype **Ll**, there would be 1000 **L** alleles and 1000 **l** alleles in the gene pool. This is a 50:50 split or, put another way, the frequency of the **L** allele would be 0.5 and the frequency of the **l** allele would be 0.5. Notice that frequencies are always expressed as decimal fractions. In the next section, we will examine allele frequencies further.

**Q** 6 What is a gene pool?

## The Hardy–Weinberg principle and the Hardy–Weinberg equation

The Hardy–Weinberg principle involves the frequencies of alleles, genotypes and phenotypes in a given population. Suppose an imaginary gene had two alleles, a dominant **A** and a recessive **a**. Three genotypes would be possible for this gene, **AA**, **Aa** and **aa**, and would be present amongst the individuals of a population. We can use symbols to represent the frequency of these alleles in the gene pool.

$p$ = the frequency of the allele **A**

$q$ = the frequency of the allele **a**          $(p + q = 1)$

Since there are only two alleles of this gene, the sum of the two frequencies must have a value of 1, i.e. the total gene pool.

**Q** 7 If the frequency of the long-wings allele (L) in a population of *Drosophila* is 0.6, what is the frequency of the vestigial-wings allele (l)?

**The Hardy–Weinberg principle** predicts that the frequencies of the alleles of a particular gene in a particular population will stay constant from generation to generation. In other words, the values of $p$ and $q$ will be the same from one generation to the next. This principle involves a number of assumptions, which you must be able to recall in a Unit test. These assumptions are as follows:

- **The population must be large** – in small populations, chance events can cause large swings in frequencies.

- **There must be no migration into the population (immigration) or out of the population (emigration)** – migration adds new alleles to, or removes alleles from, the population.

- **Mating between individuals in the population must be completely at random** – this ensures an equal chance of the alleles being passed on to the next generation.

- **No genetic mutations must occur** – gene mutations will change allele frequencies.

- **All genotypes must be equally fertile** – this ensures an equal chance of the alleles being passed on to the next generation.

We will see later in the chapter what happens if any of these conditions are not met.

**Q** 8 **What does the Hardy–Weinberg principle tell us about the frequency of alleles of a particular gene?**

We cannot tell which alleles of a gene an organism possesses. What we can see, or detect, is its phenotype. **The Hardy–Weinberg equation** enables us to calculate allele frequencies by observing phenotypes. Using the symbols that we have used so far, $p$ to represent the frequency of allele **A** and $q$ to represent the frequency of allele **a**, we can represent the frequency of genotypes in our imaginary population.

The frequency of the genotype **AA** = $p^2$

The frequency of the genotype **Aa** = $2pq$

The frequency of the genotype **aa** = $q^2$          $(p^2 + 2pq + q^2 = 1)$

You do not need to understand how these frequencies are derived or why $p^2 + 2pq + q^2 = 1$, but you must learn how to use this formula. Let's use the formula in a specific example involving wing length in *Drosophila*. In a sample of 500 fruit flies, there were found to be 480 long-winged flies and 20 vestigial-winged flies. Table 2.5 shows the steps we should always follow in calculations involving the Hardy–Weinberg equation using the values from this sample of *Drosophila*.

| Step in calculation | Calculation from example in text |
| --- | --- |
| 1. Work out the frequency of the homozygous recessives. (This is $q^2$.) | There are 20 vestigial-winged flies in the sample of 500. Their frequency is 20/500 = 0.04 |
| 2. Take the square root of the above value to get the frequency of the recessive allele ($q$) | $q^2 = 0.04$, so $q = \sqrt{0.04} = 0.2$ |
| 3. Find the frequency of the dominant allele (p) using the equation $p + q = 1$ | $p + 0.2 = 1$, so $p = 1 - 0.2 = 0.8$ |
| 4. Put these values of $p$ and $q$ into the Hardy-Weinberg equation to get the genotype frequencies | Frequency of **ll** = $q^2 = 0.04$ Frequency of **LL** = $p^2 = 0.8^2 = 0.64$ Frequency of **Ll** = $2pq = 2 \times 0.2 \times 0.8 = 0.32$ |
| 5. Check that the frequencies total 1, to make sure you are right | $0.04 + 0.64 + 0.32 = 1$, so we know our calculation is correct |

Table 2.5
The steps you should follow in calculations involving the Hardy–Weinberg equation. Note that you always start with the homozygous recessive individuals

**Q** 9 Suggest why you should always start calculations involving the Hardy-Weinberg equation with the homozygous recessive individuals rather than those showing the dominant characteristic.

## Natural selection

We have already learned in this chapter that one of the conditions of the Hardy–Weinberg principle is not met – gene mutations occur at a predictable rate. In addition we know that some populations are, or have been, very small and that mating is not always random. All of these factors will lead to changes in frequency of the alleles of a particular gene in a population.

One of the most important factors that can lead to a change in allele frequency is natural selection. **Natural selection** results from the differential fertility of organisms that, in turn, results from their genotype. In its natural environment, the size of a population is limited by a number of environmental factors. For example, animals compete for food and plants compete for space, water and access to light. Some organisms in each population will have inherited characteristics that ensure they are more successful than others in obtaining scarce resources. These successful organisms will, consequently, have more of the resources they need for growth and reproduction: they are said to be biologically **fit**. The less successful organisms will have less of the resources they need for growth and reproduction. As a consequence, they are likely to die younger or to have fewer offspring than their more successful competitors. Whether they die or simply have fewer offspring, the less successful (less fit) organisms will not pass on their alleles to the next generation as much as the fit organisms will. This causes a change in the allele frequency of that population. Notice that natural selection is only effective in changing allele frequencies if the advantageous characteristics possessed by successful organisms are inherited.

Before going further, we will look at an example of natural selection to reinforce your understanding of the stages involved. The animal in Figure 2.11 is a brown rat, which is common in Britain. Rats are a pest because they eat food intended for humans and spread disease. Since the 1950s, a substance called warfarin has been used to kill rats. Food containing warfarin is placed where rats will find it and eat it. Once in their bodies, warfarin interferes with the rats' blood clotting, so that badly poisoned rats suffer fatal haemorrhages. A few years after warfarin was first used as a pesticide against rats, resistant populations were found in parts of Britain. What seems to have happened is that a chance mutation occurred in one rat, producing a new gene allele that reduced the effectiveness of warfarin. As a result of its resistance to warfarin, this rat competed more successfully than other rats for food and mates. As a result, it produced more offspring than susceptible rats. As it passed on its allele for resistance to many of its offspring, the resistance allele became more common in the new generation. In other words, a change occurred in the allele frequency of this rat population. Eventually, most of the rats in the population became resistant to warfarin.

**Figure 2.11**
Brown rats eat human food stores and spread disease. Since 1950, warfarin has been used to kill rats in areas where they are a serious pest. Through natural selection, warfarin-resistant populations of rats have evolved

Table 2.6 summarises the stages in the evolution of warfarin-resistant populations of rats. You can use the layout of this table as a plan when answering questions about natural selection in a Unit test. Notice that, in this example of natural selection, there has been a shift in the frequency distribution of the rat population with respect to susceptibility to warfarin. We will now look at the way that natural selection changes frequency distributions.

| Sequence of events leading to natural selection | Application of these events to the evolution of warfarin resistance in populations of brown rats |
|---|---|
| 1. There is competition within the population for environmental resources that are limited | Members of a rat population compete for food |
| 2. There is genetic variation within the population | A random gene mutation caused one rat to become resistant to the pesticide, warfarin |
| 3. One genetic variant confers an advantage on the possessor(s) of a favourable allele of a gene | The rat with the allele for warfarin resistance would not suffer haemorrhages where warfarin was used, whereas a rat with only alleles for warfarin susceptibility would suffer haemorrhages. |
| 4. The possessor of the favourable allele will have more offspring than the possessor of only the less favourable alleles. | Being stronger and healthier than its susceptible competitors, the resistant rat will successfully raise more offspring. |
| 5. The possessor of the favourable allele will pass this allele on to some of its offspring. As a result, the frequency of this allele in the population will increase, i.e. natural selection has occurred. | The resistant rat passed on the resistance allele to some of its offspring. As a result, there were more resistant rats in the offspring generation. Repeated over several generations, the warfarin-resistance allele became very common in the rat population. |

Table 2.6
The evolution of warfarin resistance in populations of rats. You can apply the scheme in the left-hand column of the table to any question in a Unit test that is about natural selection

**Stabilising selection** acts against both the extremes in a range of variation. Look at Figure 2.12. The upper graph represents the frequency distribution of phenotypes in a population before stabilising selection has occurred. The graph has a fairly wide bell-shaped curve, with its mode marked in red. The lower graph has the same mode but the curve is now a much thinner bell-shape. Selection has eliminated those organisms at the extremes of the variation. This type of selection occurs on birth mass in humans. Despite medical advances, babies with very high, or very low, birth mass have a higher infant mortality rate than those at the mode for birth mass.

**Q 10** Which curve in Figure 2.12 has the smaller standard deviation?

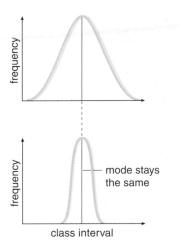

Figure 2.12
The effect of stabilising selection is to reduce variation in a population about its modal value. The upper graph represents the range of phenotypes in a population before selection and the lower graph represents the range after selection

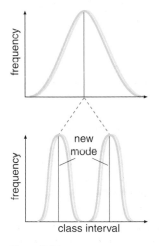

Figure 2.14
The effect of disruptive selection is to produce a bimodal distribution. Each of the new modes represents a different form (morph) of the organism that is best suited to a slightly different habitat

**Directional selection** acts against one of the extremes in a range of variation. As a result, it moves phenotypic variation away from its old modal value. This is shown in Figure 2.13. The upper and lower graphs are similar to Figure 2.12, but notice how the lower graph has shifted to the right. This type of selection has occurred in populations of the myxoma virus in Australia. This virus causes myxomatosis, a disease of rabbits, and was introduced into Australia by European settlers with the intention of killing rabbits that were pests on farms. At first, all rabbits died within a short time of infection. However, the disease is now less virulent and infected rabbits survive longer. This is thought to relate to the method of transmission of the virus from one rabbit to another. In Australia, the virus is spread by mosquitoes. Since mosquitoes will only bite a live rabbit, a myxoma virus that killed its host too quickly was likely to get stranded in a dead rabbit. Viruses that were less virulent enabled their hosts to live longer, increasing their chances of being passed to another rabbit by a biting mosquito. As a result, directional selection acted against virulent myxoma viruses and selected less virulent myxoma viruses.

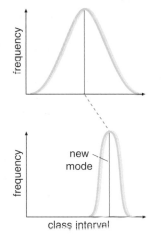

Figure 2.13
The effect of directional selection is to move the modal value. This usually occurs following a change in the environment and the new mode coincides with the new optimum value

**Disruptive selection** acts against the mode in a range of variation. Figure 2.14 shows how this type of selection produces a bimodal distribution, which might eventually result in two distinct forms of the species, called **morphs**. This type of selection occurs on the wing colour of many insects that rest during the day, when they are susceptible to predation. Figure 2.15 shows two morphs of the peppered moth, *Biston betularia*. This moth is common in Britain. It rests on tree trunks during the day, when it is preyed upon by birds, such as the one in Figure 2.16. In areas that are free of pollution, tree trunks have a blotchy pattern caused by the colour of bark and of the mosses and lichens that grow on the bark. Here, the peppered morph is less conspicuous than the black morph. As a result, birds eat more black morphs than peppered morphs and the allele for peppered wings increases in frequency. In badly polluted areas, sulphur dioxide kills lichens and mosses and soot blackens tree trunks. Here, the black morph is the less conspicuous on tree trunks. As a result, birds eat more peppered morphs than black morphs and the allele for black wings increases in frequency.

**Figure 2.15**
Two morphs of the peppered moth. Resting on tree trunks that are mottled with patches of lichen and moss, the peppered morph is less conspicuous to predatory birds than the black morph. In polluted areas, lichen and moss are killed and tree trunks become blackened by soot. Here, the black morph is less conspicuous

**Figure 2.16**
This bird has caught the conspicuous morph of the peppered moth and will eat it or feed it to its young

**Q 11  Is the evolution of warfarin resistance in rats an example of stabilising, directional or disruptive selection?**

Mutation plays an important part in natural selection, since it is a source of new genes. Where meiosis does not occur in the life cycle of a species, mutation is its only source of genetic variation. This is the case with bacteria. Many pathogenic bacteria have become resistant to the antibiotics that were used to treat infected humans. Figure 2.17 shows how this is thought to have occurred. Quite by chance, a gene mutation occurred in a single bacterial cell in a population, conferring resistance to an antibiotic. In the presence of the antibiotic, all susceptible cells in the population were killed but the resistant cell survived. Following reproduction of this cell, the entire population became resistant to the antibiotic.

**Figure 2.17**
The evolution of antibiotic-resistant populations of bacteria. A chance gene mutation enabled one bacterial cell to survive when the rest of its population was killed by antibiotic. The new population formed from this one survivor all carried the allele for antibiotic resistance

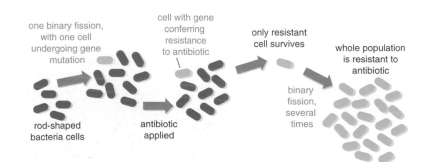

**Extension box 2**

## Balanced polymorphism in *Cepaea nemoralis*

Figure 2.18 shows different morphs of the common five-banded snail, *Cepaea nemoralis*. You can see shells of different colour, some shells with dark-brown bands and some shells without bands around their shells. The frequency of these morphs is different in different habitats, yet in each habitat they stay relatively constant from year to year. This is called a balanced polymorphism.

Shell colour can be yellow, pink or brown and is controlled by a single gene with three alleles. A different gene controls the presence or absence of bands. However, the two genes are located so close together on one of the snail's chromosomes that they form a so-called 'supergene'. The colour and banding patterns are subject to natural selection that is different in different environments. On sunny mountain sides in the Pyrenees, snails with unbanded, yellow shells absorb less heat than brown shells or banded shells. These unbanded, yellow shells are less likely to die from overheating. As a result, they breed more successfully than the other morphs and snails with unbanded, yellow shells are the most common morph.

In beech woodlands around Britain, concealment from predators is the selective force. Among the browny-pink leaf litter on the floor of a beech wood, yellow shells are conspicuous. They are easily seen by birds, such as the song thrush, which eats snails. In such woodlands, pink-shelled snails are the most common morph.

**Q 12** Bands on the shell can also affect concealment from predators. Suggest why snails with banded, yellow shells might be at an advantage over snails with unbanded, yellow shells in the mixed grass and shrubs of a hedge.

**Figure 2.18**
These shells of *Cepaea nemoralis* show variation in shell colour and in the presence or absence of bands around the shell

It is tempting to believe that human populations are unaffected by natural selection. This is not so. Figure 2.19 shows a human blood smear. The large, round cells are red blood cells. You will probably recognise them from your AS course. You are less likely to recognise the crescent-shaped, **sickle cells**, which are also red blood cells. The two types of red blood cell contain haemoglobin-A molecules that differ by a single amino acid. The alleles of the gene for haemoglobin-A are $Hb^S$ and $Hb^A$. The two alleles are codominant. Table 2.7 summarises the possible genotypes involving the $Hb^S$ and $Hb^A$ alleles of the haemophilia-A gene and the phenotypes they produce.

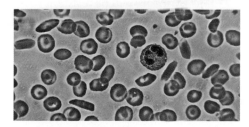

**Figure 2.19**
A human blood smear showing normal red blood cells and sickle cells

**Table 2.7**
The genotypes and phenotypes resulting from the alleles of the gene for haemoglobin-A

| Genotype | Phenotype |
|---|---|
| Hb$^A$Hb$^A$ | All red blood cells are round and carry oxygen well |
| Hb$^A$Hb$^S$ | Half the red blood cells are round and carry oxygen well; the other half carry oxygen less well and collapse in low oxygen concentration (the sickle cells in Figure 2.19). |
| Hb$^S$Hb$^S$ | All the red blood cells are sickle cells |

You might expect from Table 2.7 that individuals with either of the genotypes Hb$^A$Hb$^S$ or Hb$^S$Hb$^S$ would suffer anaemia. This is the case – they both suffer from a disease called **sickle-cell anaemia**, which is particularly severe in individuals with the genotype Hb$^S$Hb$^S$. You might then use the scheme in Table 2.6 to describe how the Hb$^S$ allele would be selected against and how its frequency in human populations would decline. This would be true but for one important fact – possession of the Hb$^S$ allele protects a person from malaria. Sickle cells are not attacked by the malarial parasite as successfully as round, red blood cells.

**Figure 2.20**
Natural selection acting on sickle-cell anaemia in humans. Whether the 'normal' or the sickle cell allele of the gene for haemoglobin-A is most common in a population is related to the incidence of malaria in the environment

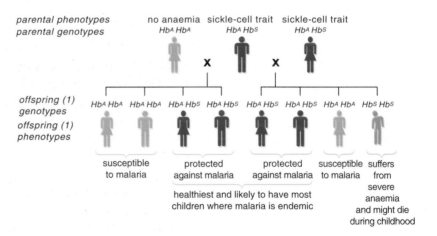

Figure 2.20 represents families with the sickle-cell condition. If one parent is Hb$^A$Hb$^A$ and the other parent is Hb$^A$Hb$^S$, there is a 50% chance that any of the offspring will be susceptible to malaria or resistant to malaria. If these individuals live in a part of the world where malaria is common, the susceptible offspring might have fewer offspring because of the effect that malaria has on them. If both parents have the Hb$^A$Hb$^S$ genotype, you will see in Figure 2.20 that there is a 75% chance that any of the offspring will be resistant to malaria. However, one of the genotypes (Hb$^S$Hb$^S$) will be so severely anaemic that she/he is unlikely to be healthy enough to have children. In this example of natural selection, it is the heterozygotes that are at an advantage in parts of the world where malaria is common.

**Q** 13 Why would you expect natural selection to reduce the frequency of the sickle-cell allele in malaria-free parts of the world?

# Summary

- When taking samples, errors can occur through chance or through human bias. We can reduce these errors by taking more than one sample and by using random sampling techniques.

- Data collected by sampling from a population can be represented as a frequency distribution. A normal frequency distribution is a special type of distribution in which the mean, median and mode have the same value.

- The standard deviation is a measure of the way that data are spread out from the mean. The more the data are spread out from the mean, the greater the standard deviation.

- The standard error (SE) is calculated from the standard deviation of a sample and is used to make allowances for the difference between samples and the total population from which they were taken. 95% of data lie within a range ± 1.96 SE of the mean

- An organism's phenotype results from an interaction between its genotype and factors in its environment. Whereas some characteristics are influenced by a single gene, others are influenced by many genes. This is called polygenic inheritance and gives rise to continuous variation.

- Variation exists between the members of any species. This is caused by processes that occur during meiosis and by gene mutation.

- Within a population, the frequency of the different alleles of a gene will stay the same from generation to generation provided the population is large, and there is no mutation, migration or natural selection.

- The frequency of alleles in a population, and the genotypes and phenotypes resulting from them, can be calculated using the Hardy-Weinberg equation. This is given as $p^2 + 2pq + q^2 = 1$, where $p$ is the frequency of the dominant allele of a gene and $q$ is the frequency of the recessive allele of the same gene.

- Natural selection occurs on populations when different genotypes have different fertilities. Fit organisms have more offspring than less fit organisms, so that succeeding generations inherit the advantageous alleles of the genes that made them successful. As a result, the frequency of the alleles of these genes changes.

- The relationship between areas where malaria is common and the frequency of sickle cell anaemia in humans, the evolution of pesticide resistance in rats and the evolution of resistance to antibiotics in bacteria are examples of natural selection in action.

# Assignment

## Rats and super-rats

This chapter contains a number of examples of how selection can lead to a change in the frequency of particular alleles in a population. One of the examples included was the evolution of warfarin resistance in a population of rats. In this assignment we will look at this example in a little more detail. You will need to use material from various parts of the specification as well as from this chapter to answer the questions. Before you start, it would be worth reading about proteins and protein structure in your notes or textbook.

When you cut yourself, blood clots rapidly and seals the wound. The mechanism governing blood clotting is complex and involves a number of substances called clotting factors. Some of these factors are formed in the liver and contain an unusual amino acid residue called $\gamma$-carboxyglutamate. Vitamins are substances which are needed in small amounts in the diet. They are needed because the body is unable to synthesise them. One of these vitamins is called vitamin K. (It is called this because it was discovered in Germany and named the Koagulation-vitamin.) Vitamin K is necessary for the synthesis of $\gamma$-carboxyglutamate. Figure 2.21 summarises the biochemical pathway involved. Look at this diagram carefully and then answer questions 1 and 2.

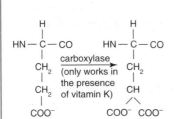

**Figure 2.21**
The biochemical reaction in which $\gamma$-carboxyglutamate is produced in the body. This reaction requires an enzyme, carboxylase, which will only work in the presence of vitamin K

**Figure 2.22**
Rats are serious pests on farms. With a lot of grain and animal food available large populations build up rapidly. Warfarin is a very effective poison and still widely used to control rats

1   Amino acids join together to form a polypeptide chain. When an amino acid has joined to another in this way it is called an amino acid residue.

   (a)  Draw a diagram to show the structure of a glutamate molecule.

*(1 mark)*

   (b)  Explain why the structure of a glutamate residue differs from the structure of a glutamate molecule.

*(2 marks)*

   (c)  Explain why the enzyme controlling this reaction is known as a carboxylase.

*(1 mark)*

2   The R-group of the $\gamma$-carboxyglutamate residue binds readily to phospholipids. Suggest why this might be useful in blood clotting.

*(2 marks)*

Warfarin is a very effective rat poison. It is a competitive inhibitor of vitamin K.

3   (a)  Look at Figure 2.22. The rat in this photograph has recently eaten warfarin. Use the information given so far in this assignment to suggest an explanation for the blood on the rat's face.

*(4 marks)*

(b) If humans are accidentally poisoned by warfarin, they are treated with large doses of vitamin K. Use your knowledge of competitive inhibition to explain why this treatment is successful.

*(1 mark)*

In 1959, rats resistant to warfarin were found in an area in central Wales. It was found that this resistance was determined by a single gene with two codominant alleles, $W^R$ and $W^S$. The three different genotypes have the following characteristics.

- $W^SW^S$ Rats with this genotype are susceptible to warfarin. They only require small amounts of vitamin K in their diet.

- $W^SW^R$ These rats are resistant to warfarin. They require more vitamin K in their diet than rats with the genotype $W^SW^S$.

- $W^RW^R$ Rats with this genotype are also resistant to warfarin but as they need extremely large amounts of vitamin K in their diets, they cannot survive under natural conditions.

4 Giving a reason for your answer in each case, which of these three genotypes would be at the greatest advantage in an area where:

(a) no warfarin had ever been used;

*(2 marks)*

(b) warfarin was commonly used as a rat poison.

*(2 marks)*

5 Use genetic diagrams to show the genotypes of the expected offspring of crosses between rats with the following genotypes:

(a) $W^SW^R$ and $W^SW^R$

(b) $W^SW^R$ and $W^SW^S$.

*(3 marks)*

6 The percentage of rats resistant to warfarin in central Wales rose rapidly at first. It then levelled out. Use the results of your answer to question 5 to explain this.

*(4 marks)*

Thrombosis is the formation of a blood clot inside a blood vessel. It can be very serious because the clot might block the flow of blood to the heart or the brain and cause the death of the patient. Patients with thrombosis are treated with anticoagulant drugs. One commonly used anticoagulant is warfarin.

7 A gene has been described which controls warfarin resistance in humans. In this gene, the allele for warfarin resistance is dominant. Suggest why this allele is rare in the human population.

*(3 marks)*

# Examination questions

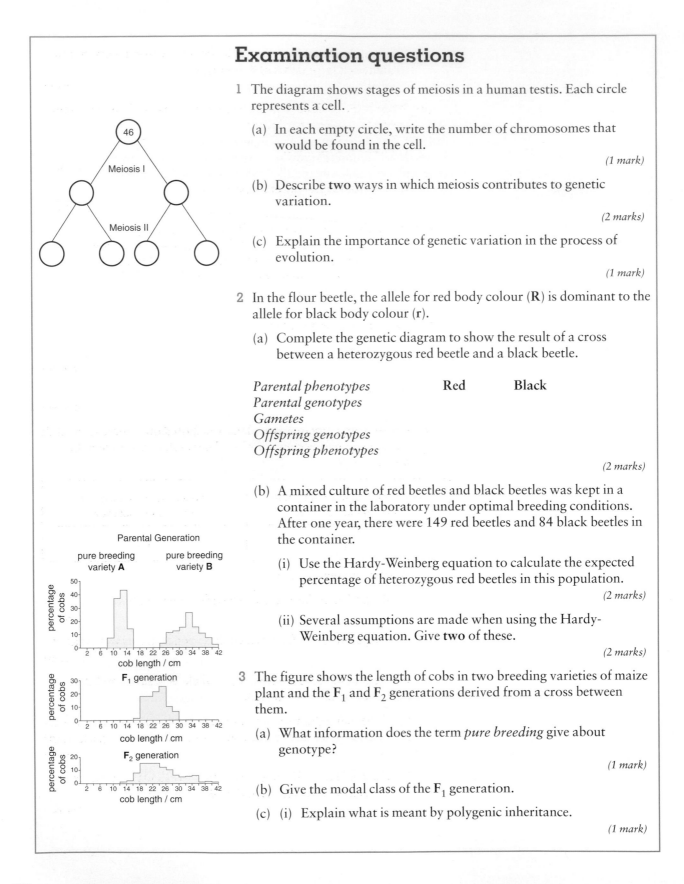

1 The diagram shows stages of meiosis in a human testis. Each circle represents a cell.

(a) In each empty circle, write the number of chromosomes that would be found in the cell.

*(1 mark)*

(b) Describe **two** ways in which meiosis contributes to genetic variation.

*(2 marks)*

(c) Explain the importance of genetic variation in the process of evolution.

*(1 mark)*

2 In the flour beetle, the allele for red body colour (**R**) is dominant to the allele for black body colour (**r**).

(a) Complete the genetic diagram to show the result of a cross between a heterozygous red beetle and a black beetle.

*Parental phenotypes*           **Red**       **Black**
*Parental genotypes*
*Gametes*
*Offspring genotypes*
*Offspring phenotypes*

*(2 marks)*

(b) A mixed culture of red beetles and black beetles was kept in a container in the laboratory under optimal breeding conditions. After one year, there were 149 red beetles and 84 black beetles in the container.

(i) Use the Hardy-Weinberg equation to calculate the expected percentage of heterozygous red beetles in this population.

*(2 marks)*

(ii) Several assumptions are made when using the Hardy-Weinberg equation. Give **two** of these.

*(2 marks)*

3 The figure shows the length of cobs in two breeding varieties of maize plant and the $F_1$ and $F_2$ generations derived from a cross between them.

(a) What information does the term *pure breeding* give about genotype?

*(1 mark)*

(b) Give the modal class of the $F_1$ generation.

(c) (i) Explain what is meant by polygenic inheritance.

*(1 mark)*

x

(ii) Explain the evidence from the figure which suggests that the inheritance of cob length is polygenic.

*(2 marks)*

(d) What is the evidence that differences in cob length in the parental generation are partly due to:

    (i) genetic difference;

*(1 mark)*

    (ii) environmental differences?

*(1 mark)*

In maize a single gene with two alleles controls the type of carbohydrate stored in the cells of the plant. Starchy varieties of maize have starch grains which stain blue-black with iodine solution; waxy varieties have starch grains which stain red. The allele for starch, **W**, is dominant to that for waxy, **w**.

(e) Explain what is meant by;

    (i) a gene;

*(1 mark)*

    (ii) an allele.

*(1 mark)*

(f) Pollen from a single maize plant was dusted on a microscope slide and stained with iodine solution. The results are shown in the table.

| Pollen grains stained blue-black with iodine solution | Pollen grains stained red with iodine solution |
| --- | --- |
| 58 | 64 |

What is the genotype of:

    (i) the pollen grains stained red with iodine solution;

*(1 mark)*

    (ii) the parent plant from which these pollen grains were taken?

*(1 mark)*

(g) In a field of maize the frequency of allele **W** was 0.7 and the frequency of allele **w** was 0.3.

    (i) If the maize plants were randomly fertilised, what frequencies of these two alleles would be expected in the next generation?

*(1 mark)*

    (ii) Use the Hardy–Weinberg equation to calculate the percentage of heterozygous plants in the field of maize.

*(3 marks)*

# 3

# The Evolution and Classification of Species

**Figure 3.1**
This bandro lives in the papyrus reed beds of Lake Alaotra in Madagascar. Biologists dispute whether bandros belong to an individual species or whether they are a subgroup of another species of lemur

The animal in Figure 3.1 is a type of lemur, called the bandro. It is found in only one habitat in the world – the papyrus reed beds around Lake Alaotra in Madagascar – where no other lemurs live.

Bandros live in small family groups. The head and body length of an adult is about 30 cm and an adult weighs about 900 g. Bandros are active during the day. They spend most of this time feeding on papyrus. Like the animal in Figure 3.1, bandros use their hands, feet and tail to hold on to the papyrus reeds. Unfortunately, bandros are classed as a critically endangered species because their habitat is being destroyed. They are also hunted for food and for use as pets.

The bandro is a confusing animal. Firstly, it has several common names – the bandro, the Lake Alaotra bamboo lemur and the grey gentle lemur. It even has two biological names – *Hapalemur alaotrensis* and *Hapalemur griseus alaotrensis*. Secondly, biologists dispute whether the bandro is a species or whether it is a localised subgroup of another species. The biological names reflect this dispute. The name *Hapalemur alaotrensis* gives bandros the status of a species whereas the name *Hapalemur griseus alaotrensis* suggests that bandros are a subspecies of the lesser bamboo lemur (*Hapalemur griseus*).

In this chapter, you will learn how biologists recognise individual species and begin to understand why there is a dispute about the status of the bandro. You will also learn how species are thought to have evolved, how they are classified and the origin of their biological names.

## The concept of a species

**Figure 3.2**
Karela is the unripe fruit of *Momordica charantia*. It is also known around the world as bitter gourd, bitter melon, balsam pear or foogwa. It is a popular vegetable in northern India, where it is believed to be good for the heart, and has been recommended by the Department of Health in the Philippines as a remedy for diabetes mellitus. However, many people find the bitter taste of karela, especially strong in its seeds, makes it unpleasant to eat

Humans seem to have a need to classify objects or organisms in the world around them. If you watch a young child putting things into its mouth, you can see it classify objects as pleasant to taste or distasteful. You continue to do this as an adult – some foods you like to eat and others you do not (Figure 3.2). You probably classify animals as dangerous or harmless, berries and fungi as poisonous or edible, people as helpful or unfriendly, and so on. When we make a formal study of different organisms, we need to agree a classification system that is universal.

**Figure 3.3**
The horse (*Equus equus*) and donkey (*Equus hemionus*) belong to different species. There are clear differences in their external appearance and, although they can interbreed, their offspring (mules) are always sterile

The **species** is the basic unit of biological classification. We can define the meaning of species in two ways.

1. **Variation** – Members of a species are similar to each other but different from members of other species. These similarities might be physical (e.g. two trees have a similar branching pattern), biochemical (e.g. the haemoglobin structure of two animals is identical), immunological (e.g. an antibody made against antigens from one organism is equally effective against those from a second organism), developmental (e.g. growth of the embryo is similar in two organisms), or ecological (e.g. organisms occupy an identical ecological niche). In Chapter 2, we learned about variation within a species and you can see this variation if you look around at your fellow students or at the members of your family. Consequently, we need to be careful about the importance of the differences that we use to identify different species.

2. **Potential for breeding** – Members of a species breed together in their own environment and produce offspring that are fertile. Look at the horse and donkey in Figure 3.3. You can see the differences between horses and donkeys that led to them being classed as different species. In captivity, they can mate and produce offspring, called mules. However, mules are always sterile. In this case, the potential for breeding confirms our classification of horses and donkeys that was based on observable differences.

These two definitions of the term species are not always helpful to biologists. Some species seldom, if ever, reproduce sexually. Other species are known only from their fossil record. Finally, some groups of similar organisms live so far from each other that they are unlikely ever to meet. In these cases, it is difficult to use our second definition to decide whether these organisms belong to a separate species.

**Q** 1 **Use the information above to suggest why biologists are undecided about the biological status of the bandro in Figure 3.1.**

## Speciation

Speciation is the formation of new species from existing species. It is often called evolution. The evolution of new species is a controversial idea, which many groups of people reject. Most biologists believe that the species we see around us have evolved from other species and this idea is central to biological classification.

New species are thought to arise in one of two ways. One of these involves:

- **hybridisation** – the production of offspring from parents of two different species
- **polyploidy** – an increase in the number of sets of chromosomes.

We saw one example of hybridisation above. Horses and donkeys can mate to produce mules, but these mules are always sterile. Their sterility is explained by the chromosomes that mules inherit from their parents.

The parental horse chromosomes and the parental donkey chromosomes are so different that they cannot form homologous pairs. Although this has no effect on mitosis, it prevents meiosis, which is essential for egg and sperm formation in animals.

**Q  2  At which stage of meiosis do homologous chromosomes pair together?**

The sterility of mules could theoretically be overcome if the mule were to double its diploid chromosome number, i.e. become **tetraploid** (4n). There would now be two copies of every individual chromosome and these could form homologous pairs during meiosis. This process has not happened in mules and rarely happens at all in animals. It is, however, common in plants. In fact, over 50% of flowering plants are polyploid. A few examples of familiar, polyploid plants are shown in Table 3.1.

| Wild species of plant | Cultivated species of plant |
|---|---|
| Wild cotton (2n = 26) | Cultivated cotton (2n = 52) |
| Wild dahlia (2n = 32) | Garden dahlia (2n = 64) |
| Wild potato (2n = 24) | Cultivated potato (2n = 48) |
| Wild rose (2n = 14) | Garden rose (2n = 42) |
| Wild tobacco (2n = 24) | Cultivated tobacco (2n = 48) |

Table 3.1
Like over 50% of flowering plants, these cultivated plants are polyploid. The diploid number of chromosomes is shown as 2n

**Extension box 1**

## Polyploidy in wheat

Wheat is a crop plant that is grown all over the world. It is used to make bread and to make pasta. Modern wheat is thought to have arisen thousands of years ago by the accidental crossing of different species of grass to produce hybrid offspring that became polyploid. Figure 3.4 shows how the evolution of modern wheat is thought to have occurred.

Figure 3.4
Modern wheat is thought to have arisen following chance hybridisation and polyploidy on at least two occasions in the past four thousand years

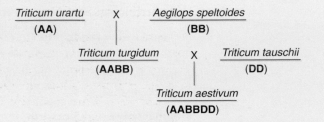

*Triticum urartu* was an ancestral wheat species that accidentally crossed with a wild grass, called *Aegilops speltoides*. The hybrid offspring of this cross would have been sterile except that a chance mutation caused it to become polyploid. This is represented in Figure 3.4, which shows the chromosome combination of the parents as **AA** and **BB**, where **A** and **B** represent one haploid set of parental chromosomes. Instead of being **AB**, the polyploid hybrid is **AABB**, i.e. it has two sets of each of the parent's haploid chromosomes. This new species is *Triticum turgidum*, commonly called durum wheat, and is used to make pasta.

A second hybridisation is then thought to have occurred in which *Triticum turgidum* formed a hybrid with *Triticum tauschii* (chromosome combination **DD**). Chance mutation of the hybrid formed a polyploid, *Triticum aestivum*. This is commonly called bread wheat because its flour is used to make bread.

Hybridisation and polyploidy can explain speciation in plants but not in animals. We need another explanation of the way in which new species are formed. This method of speciation involves:

- the separation of two groups of organisms of one species

- reproductive isolation between the two separated groups, i.e. members of the two groups cannot breed with each other, so that they are genetically isolated

- changes in the frequency of alleles that are different in the two reproductively isolated groups and that are so great that the two groups become new species

Table 3.2 (see page 58) gives some examples of the barriers that can lead to reproductive isolation. It is important to realise that, whilst reproductive isolation is likely to result in genetic differences between the isolated groups, it will not always lead to new species. For example, humans in the Australian subcontinent were reproductively isolated from humans in Asia and Europe for thousands of years. As a result, modern Australian aborigines have features that distinguish them from humans inhabiting Asia and Europe. However, breeding amongst all these human groups occurs and results in fertile offspring. All these humans still belong to the same species, *Homo sapiens*.

## Allopatric speciation

It is easiest to imagine populations diverging from each other if barriers that they cannot cross physically separate them. In their different environments, the populations will be subjected to different natural selection pressures, which will cause their allele frequencies to change in different ways. When speciation occurs as a result of such barriers, we call this **allopatric speciation**.

The Hawaiian Islands were formed thousands of years ago by the eruption of volcanoes on the floor of the Pacific Ocean. As the volcanoes cooled, they became suitable for plant and animal life. The islands were

| Stage of life cycle at which barrier is effective | Barrier to interbreeding | Explanation |
|---|---|---|
| Before mating | geographical separation | populations inhabit different continents, different islands or different sides of a large canyon |
| | habitat isolation | populations inhabit different local habitats within one environment |
| | temporal isolation | populations use the same environment but are reproductively active at different times |
| After mating | incompatibility of gametes | female might kill sperm of wrong genotype or pollen of wrong genotype fails to germinate |
| | hybrid not viable | hybrid fails to develop to maturity |
| | hybrid sterility | hybrid grows to maturity but is sterile (Figure 3.3) |

**Table 3.2**
Examples of barriers that lead to reproductive isolation of two populations of a single species. If sufficient genetic differences occur in the isolated populations, new species are formed

probably colonised by organisms from the neighbouring mainland, in a random way. The only group of flies that colonised the Hawaiian Islands was the fruit fly, *Drosophila*. There are now hundreds of species of fruit fly in these islands. They are thought to have evolved from an ancestral population through allopatric speciation.

**Q** 3 **Name the barrier(s) to reproduction that might have led to the evolution of different species of fruit fly in the islands of the Hawaiian archipelago.**

## Sympatric speciation

The formation of new species without geographical isolation is called **sympatric speciation**. Where this occurs, strong forces of natural selection cause genetic differences between two populations that can, at least in theory, interbreed.

The waste from copper mines forms unsightly tips in many parts of Britain. One species of grass, called *Agrostis tenuis* (common bent grass), is able to grow on these tips because some plants possess an allele of a gene that makes them tolerant to high concentrations of copper ions in the soil. These plants can grow on polluted soil that is inhospitable to susceptible *A. tenuis* plants. Sometimes, the copper-tolerant populations grow very close to populations of *A. tenuis* that are susceptible to poisoning by high concentrations of copper ions. However, copper-tolerant plants have an earlier flowering time than the copper-susceptible plants. Although these two varieties of *A. tenuis* belong to the same species, there is a potential for reproductive isolation that might lead to sympatric speciation.

**Q** 4 **Distinguish between allopatric speciation and sympatric speciation.**

## The five-kingdom classification of organisms

In our biological classification system, all species are placed in larger groups, called genera (singular: genus). A single **genus** contains species that, although different from each other, have similarities that make them distinct from other genera. In turn, genera are placed into larger groups, and so on. Figure 3.5 shows the names of these biological groups, called **categories**. Notice that organisms in the categories at the bottom of Figure 3.5 show a greater degree of similarity than those in categories at the top of the diagram. Each set of organisms within a category is called a **taxon**. Taxonomy is the study of biological classification.

Table 3.3 shows how three animals previously named in this chapter are classified. It also shows the classification of a familiar plant, the tomato. Each name in the body of the table is a taxon. The names are in Latin, which no one uses today and so the names cannot be corrupted. The taxa (plural of taxon) shared by humans, horses and fruit flies are shown in green. Notice that the biological name of an organism is the **binomial** formed from the names of its genus and species. Through international convention, the name of the genus is always written with an upper case and the name of the species is always written with a lower case. The binomial is always printed in italics (e.g. *Homo sapiens*) or, if written by hand, is underlined (e.g. Homo sapiens).

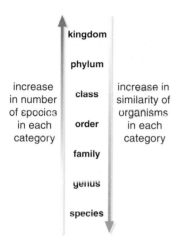

**Figure 3.5**
Species are grouped into biological categories of decreasing similarity. The largest category is the biological kingdom, of which there are five

| Category | Human | Horse | Fruit fly | Tomato |
|---|---|---|---|---|
| kingdom | Animalia | Animalia | Animalia | Plantae |
| phylum | Chordata | Chordata | Arthropoda | Angiospermophyta |
| class | Mammalia | Mammalia | Insecta | Dicotyledoneae |
| order | Primata | Perissodactyla | Diptera | Solanales |
| family | Hominidae | Equidae | Drosophilidae | Solanaceae |
| genus | *Homo* | *Equus* | *Drosophila* | *Lycopersicon* |
| species | *sapiens* | *equus* | *melanogaster* | *esculentum* |

**Table 3.3**
The taxa of three animals, mentioned earlier in this chapter, and of a familiar plant, demonstrate the major categories used in classification. Humans, horses and fruit flies share the taxa shown in green

The first three organisms in Table 3.3 belong to the same kingdom – Animalia. This is one of only five kingdoms into which all organisms are grouped. The five kingdoms are:

- **Prokaryotae** (prokaryotes, which are all bacteria)
- **Protoctista** (protoctists)
- **Fungi** (fungi)
- **Plantae** (plants)
- **Animalia** (animals)

We will now look at the distinguishing characteristics of each kingdom, which are summarised in Table 3.4. In a Unit test, you will not be expected to recall the classification of any specific organism.

**Q** 5 **What is the binomial for the tomato plant?**

**Table 3.4**
A simple comparison of the five kingdoms. Unicellular organisms consist of only a single cell, colonial organisms are groups of cells and multicellular organisms have many cells arranged in tissues. Autotrophic organisms use external energy sources to form organic compounds from inorganic compounds whereas heterotrophic organisms digest organic molecules into smaller products, which they then absorb into their own bodies

| Kingdom | Cell structure | Cell wall | Nutrition | Notes |
|---|---|---|---|---|
| Prokaryotae | prokaryotic; unicellular or colonial | present (peptidoglycans) | some autotrophic (photosynthesis and chemosynthesis), some heterotrophic | contains only the bacteria |
| Protoctista | eukaryotic; includes unicellular, colonial and multicellular forms | sometimes present (polysaccharide) | some autotrophic, some heterotrophic, some both | organisms are classed in this kingdom if they cannot be placed in any other |
| Fungi | eukaryotic; can be single-celled (e.g. yeast) but most are multicellular | present (chitin) | heterotrophic | most fungi are made of a mass (mycelium) of thread-like filaments, called hyphae; they lack cilia and flagella at all stages of their life cycle; they reproduce by forming resistant spores, which they produce by mitosis |
| Plantae | eukaryotic and multicellular | present (cellulose) | autotrophic (photosynthesis using chloroplasts) | plants develop from multicellular embryos, which are formed from zygotes and nourished by the maternal plant; they have a complex life cycle that involves two different types of adult – a haploid, sexually reproducing gametophyte and a diploid, asexually reproducing sporophyte |
| Animalia | eukaryotic and multicellular | absent | heterotrophic, involving a digestive cavity | animals develop from embryos that, at some stage, form a hollow ball of cells – the blastula; they have nervous and hormonal control systems |

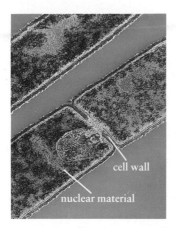

cell wall

nuclear material

## Prokaryotae

In your AS course, you learned about prokaryotic and eukaryotic cells (e.g. refer back to pages 14 and 15 of the AS book in this series). Table 3.5 summarises the differences between prokaryotic cells and eukaryotic cells. All organisms that have prokaryotic cells are classed in the kingdom Prokaryotae. They are all bacteria, such as the bacterium shown in Figure 3.6. There are about 10 000 known species of bacteria.

**Figure 3.6**
An electron micrograph of a rod-shaped bacterium (bacillus)

| Feature | Prokaryotic cells | Eukaryotic cells |
|---|---|---|
| size | relatively small, of the order of 1 to 10 μm | relatively large, of the order of 10 to 100 μm |
| cell structure | cell capsule sometimes present | cell capsule never present |
| | cell wall usually present and made of polymers of sugars and amino acids (peptidoglycans) | cell wall present in some protoctists (cellulose), fungi (chitin) and plants (cellulose) |
| | cytoplasm lacks membrane-bound organelles | cytoplasm contains membrane-bound organelles, such as mitochondria |
| | ribosomes relatively small (70S*) | ribosomes relatively large (80S*) |
| genetic material | DNA not coated by proteins | DNA coated by proteins (called histones) |
| | DNA forms a single circular genophore | DNA forms one, or more, linear chromosomes |
| | DNA free in cytoplasm as a nucleoid | chromosomes held within a membrane-bound nucleus |
| cell division | neither mitosis nor meiosis is involved | division is by mitosis or meiosis |
| | cytoplasm divides into two (binary fission) and replicated genophore is separated without a spindle of microtubules | cytoplasm divides into two and replicate chromosomes are separated by a system of microtubules, which form a spindle |
| cell metabolism | great variation in metabolic pathways | similar metabolic pathways |
| | some photosynthesise, forming sulphur, sulphates or oxygen as a waste product | some protoctists and most plants photosynthesise and form oxygen as a waste product |
| | include obligate aerobes (need free oxygen for respiration), obligate anaerobes (need absence of free oxygen for respiration) and facultative aerobes (can adapt their respiration to both aerobic and anaerobic conditions) | most are obligate aerobes, though some fungi are facultative aerobes |

**Table 3.5**
A comparison of prokaryotic and eukaryotic cells. All bacteria have prokaryotic cells and belong to the kingdom Prokaryotae. (*) The symbol S stands for Svedberg unit and is a measure of sedimentation rate during centrifugation. A 70S ribosome has a slower sedimentation rate than an 80S ribosome because it is smaller

## Protoctista

This kingdom contains about thirty different phyla (plural of phylum), whose members are all simple-bodied eukaryotes. There are about 100 000 known species of protoctists. The kingdom Protoctista is best described by exclusion, i.e. if an organism is not a prokaryote, a fungus, a plant or an animal, then it is a proctoctist. Among the protoctists are:

- organisms that consist of only one cell (**unicellular**), organisms that form filaments of cells, organisms that form ball-like colonies of cells and organisms that are made of many cells organised into tissues (**multicellular**)

- organisms that behave like animals (e.g. *Plasmodium*, which causes malaria) and organisms that behave like plants (e.g. seaweeds)

- slime moulds, protozoa and algae

Figure 3.7 shows examples of all these types of protoctists. Please note that the terms protozoa (unicellular, 'first animals') and algae (unicellular or very simple multicellular photosynthetic organisms), though commonly used, do not refer to taxa.

**Figure 3.7**
Examples of protoctists
(a) *Plasmodium* is an animal-like, single-celled organism that causes malaria in humans (×400)
(b) *Spirogyra* is a filament of photosynthetic cells (×400)
(c) Volvox is a ball-like colony of photosynthetic cells (×40)
(d) The seaweed wrack, *Fucus*, is a photosynthetic, multicellular protoctist

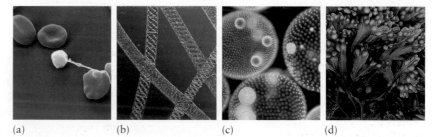

(a)  (b)  (c)  (d)

## Extension box 2

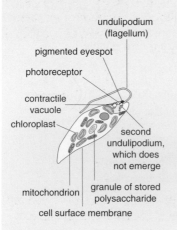

undulipodium (flagellum)
pigmented eyespot
photoreceptor
contractile vacuole
chloroplast
second undulipodium, which does not emerge
mitochondrion
granule of stored polysaccharide
cell surface membrane

**Figure 3.8**
*Euglena* is a protoctist that has some plant-like qualities and some animal-like qualities

## *Euglena* – a protoctist with both animal and plant-like qualities

Figure 3.8 shows a protoctist called *Euglena*, which is common in freshwater ponds. You can see that it is a unicellular eukaryote with many chloroplasts. When a single *Euglena* is in light, these chloroplasts photosynthesise in just the same way as those in plants. For this reason, you might be tempted to classify *Euglena* as a plant. However, you can see that *Euglena* also has properties that are not shared by plants. For example, it has no cellulose cell wall. As a result:

- it is not able to prevent the inward movement of water by osmosis. The contractile vacuole that you can see in the diagram uses energy from ATP to collect this water and pump it out of the cell, preventing osmotic lysis of the cell.

- it is able to change its shape.

Unlike a plant, *Euglena* is able to move about. Waves of contractions pass down its flagellum, in a whip-like action that pulls the cell forwards and makes it spin at the same time. Contractile fibrils in its outer surface change the shape of the cell and can also move it forwards, in a process called euglenoid movement. The photoreceptor, which you can see at the base of the flagellum in the diagram, helps *Euglena* to detect light. It moves towards light of moderate intensity (positive phototaxis) but away from very bright light (negative phototaxis).

Although *Euglena* can photosynthesise, in darkness it can feed like fungi and many bacteria, i.e. by secreting enzymes onto organic molecules in its surroundings and absorbing the digested products.

Our inability to classify *Euglena* clearly into one of the other kingdoms results in it being classified in the kingdom Protoctista.

**Q** 6 **How can you tell from Figure 3.8 that *Euglena* has a eukaryotic cell?**

**Figure 3.9**
Like all fungi, this puffball produces fruiting bodies, which is usually all we see of a fungus. The fruiting bodies are made of masses of thread-like hyphae and contain the reproductive spores that all fungi produce. You can see some of the spores of this puffball being released in a smoke-like cloud

**Figure 3.10**
Fungi reproduce asexually and sexually to form spores. On a suitable environment, a single spore germinates to form a fungal hypha. Continued growth and branching forms a network of hyphae, or mycelium

## Fungi

Fungi are commonly called moulds because we see them most often on objects that are decaying. If you have kept fruit or bread for too long, you will have seen it go 'mouldy'. The mould that you see is mainly the fruiting bodies of the fungi. These fruiting bodies form spores, which is the way in which fungi disperse. If you look at Figure 3.9, you will see spores being released into the air by the fruiting body of a puffball fungus. Like mushrooms, many puffballs are fungal fruiting bodies that are eaten by humans.

There are about 100 000 known species of fungi. They all have eukaryotic cells, which are surrounded by a cell wall made of chitin. Some fungi are unicellular (e.g. yeast), but most have bodies that are made of thread-like filaments, called hyphae (singular, **hypha**). Once a fungal spore lands on a suitable environment, it germinates to form a single hypha. Figure 3.10 shows how, by continued growth and branching, a network of hyphae, called a **mycelium**, is formed. This is the body form of the great majority of fungi. Although you cannot see them with the naked eye, vast fungal mycelial networks surround us in soil and anywhere where there is decaying organic matter.

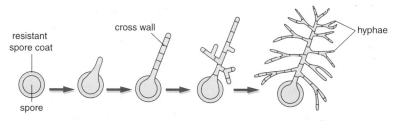

Figure 3.11(a) shows a section through a hypha. In it, you can see the cell wall. This is made of chitin and, in some fungi, separates the hypha into cells. A hypha is most active at its tip. It is here that:

● growth of the hypha occurs

● most digestive enzymes are released onto the surroundings

● most absorption of digested food back into the hypha occurs.

In contrast, Figure 3.11(b) shows a section through a yeast cell. This unicellular fungus is used in industry to make beer and wine and to make bread rise. Cultures of yeast are commonly used in school and college laboratories to investigate respiration.

**Figure 3.11**
(a) The hypha shown in this diagrammatic section has walls across its length, which are not present in all fungi (b) Yeast is a unicellular fungus. During aerobic respiration, cultures of yeast produce large amounts of carbon dioxide gas: this causes bread to rise in the baking industry. During anaerobic respiration, cultures of yeast produce ethanol: this is the basis of the beer and wine industries

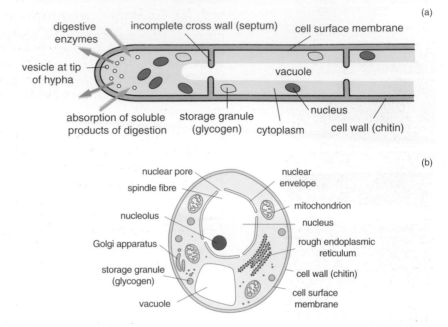

**Q 7 Suggest why there are usually many nuclei concentrated in the cytoplasm at the tip of a fungal hypha.**

The mould that you might have seen on old bread was mentioned above. One of the fungi that is most likely to have caused this mould is a species of *Penicillium*. Figure 3.12 shows the generalised structure of fungi in the genus *Penicillium*. In this diagram, you can see most of the features that characterise fungi. You can see:

- a mycelium, formed by hyphae that are separated into cells by cross walls

- special feeding hyphae, called haustoria, that secrete digestive enzymes onto the surroundings and absorb the products of digestion

- erect hyphae that grow upwards from the mycelium

- chains of spores on the erect hyphae. It is the colour of these spores that you would notice on mouldy bread. It is also the brush-like appearance of these chains of spores that gave the genus *Penicillium* its name, from the Latin penicillus, or little brush.

*Penicillium* can be a nuisance. Apart from causing stored food to go mouldy, it also causes spoilage of stored textiles. However, *Penicillium* does have its uses. Many species of *Penicillium* live in soil, where they contribute to nutrient cycles by breaking down dead organic matter. *P. roqueforti* and *P. camemberti* are used to ripen cheeses. *P. notatum* is the species in which Sir Alexander Fleming first discovered the antibiotic penicillin.

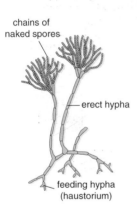

**Figure 3.12**
The general features of the genus *Penicillium*. The body of this fungus is a mass of hyphae which grow horizontally through the medium on which it lives. Erect hyphae grow from this mycelium and hold spores, which are dispersed by touch or air currents. The spores are blue, green or yellow and give rise to the coloured mould you might see on stored food

(a)

(b)

(c)

(d)

(e)

**Figure 3.13**
These organisms are representative of five different plant phyla (a) Liverwort (b) Moss (c) Fern (d) Conifer (e) Flowering plant

**Figure 3.14**
Plants have a complex life cycle, known as alternation of generations because it involves two separate types of adult. Sexual reproduction by gametophyte adults forms multicellular zygotes that develop into sporophyte adults. These sporophytes reproduce asexually to form spores that germinate into gametophytes

## Plantae

There about 350 000 known species of plants. All plants are multicellular organisms, with eukaryotic cells containing large vacuoles and surrounded by cellulose cell walls. They usually grow into a branched body form and grow from restricted patches of actively dividing cells, called meristems. Most plants photosynthesise, using chloroplasts, and produce oxygen as a waste product of their photosynthesis. Figure 3.13 shows a variety of plants. The liverwort and moss are relatively small, so that you might not notice them during a normal day. Ferns, conifers and flowering plants are large and you are likely to see examples of these plants every day.

Plants are adapted to living on land and many of their features are adaptations to life on land. They have strong supporting tissues, their leaves allow gas exchange in the air and they are fairly well waterproofed to prevent drying out. The life cycle of plants is complex and includes two different types of adults. Figure 3.14 summarises this life cycle, known as alternation of generations. The parts of the liverwort, moss, fern, conifer and flowering plants shown in Figure 3.13 are spore-producing adults (sporophytes). Their gamete-producing adults are hidden within them and you would not see them without an optical microscope.

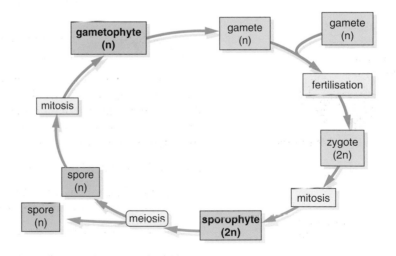

## Extension box 3

## Alternation of generations in flowering plants (phylum, Angiospermaphyta)

Alternation of generations is a feature of the plant kingdom. In its life cycle, every plant has two different types of adult: a sexually reproducing gametophyte and an asexually reproducing sporophyte. These are difficult to distinguish in flowering plants because the gametophytes are kept within the flowers produced by the sporophyte.

**Figure 3.15**
Alternation of generations in a flowering plant. The sexually reproducing gametophytes are haploid and are kept within the body of the diploid, asexually reproducing sporophyte

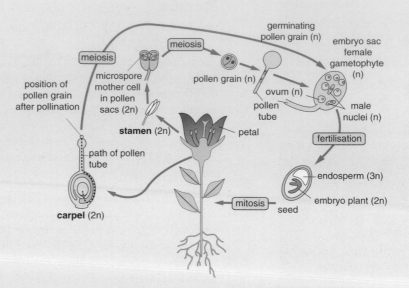

Figure 3.15 represents the life cycle of a flowering plant. What we would normally call the plant is actually the diploid sporophyte generation. In its flowers, it forms two types of haploid gametophyte, a male (the germinated pollen grain) and a female (the embryo sac). The embryo sac contains a number of haploid nuclei. One of these is the egg cell. It is fertilised by one haploid nucleus from the germinated pollen grain, which travels down the pollen tube that grows from the pollen grain to the embryo sac within the flower. This fertilisation forms a zygote, which divides by mitosis to form an embryo plant. A second haploid nucleus in the pollen tube fuses with two of the other haploid nuclei within the embryo sac to form a triploid nucleus (3n). This divides by mitosis to form the endosperm that nourishes the growing embryo plant. The embryo sac and endosperm are enclosed with a protective seed.

Plants are incredibly useful to humans. They provide us with food, building materials, rubber and many drugs that are used in the pharmaceutical industry.

## Animalia

Animals are multicellular organisms, whose cells are eukaryotic and always lack cell walls. Animals are unable to photosynthesise and so are always heterotrophic. They have a variety of ways of obtaining nourishment from ready-made organic sources (food) and a variety of body cavities for digesting and absorbing inorganic nutrients from their food. You will learn more about the structure of the digestive system of humans in Chapter 12. Animals develop from a multicellular embryo that, at some stage, forms a hollow ball of cells, called a **blastula**.

The animal kingdom contains over 1 million species, which are organised into many different phyla. They include sponges, jellyfish, corals, worms, crabs, insects, snails, squid, sea urchins, fish, amphibians, reptiles, birds and mammals. Figure 3.16 shows examples of some of these phyla. As we saw in Table 3.3, we are animals, belonging to the phylum Chordata.

**Figure 3.16**
The animal kingdom contains a huge variety of organisms. The distinctive feature of them all is that they developed from a multicellular embryo that formed a hollow ball of cells, called a blastula (a) Jellyfish (b) Earthworm (c) Insect (d) Slug (e) Starfish (f) Fish (g) Reptile (h) Bird (i) Mammal

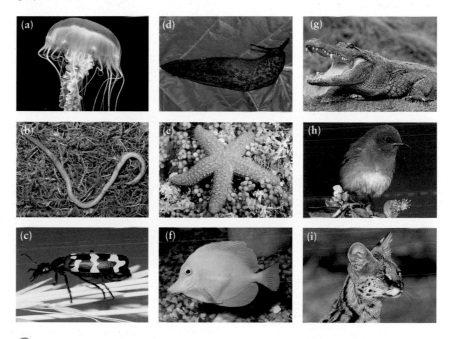

**Q** 8 An organism has cells, which contain nuclei and are surrounded by cell walls. To which of the five biological kingdoms might this organism belong?

## Summary

- A species is the smallest unit of biological classification.

- Species can be described in two ways.

  1. Variation – members of the same species are very similar to each other but different from members of other species.

  2. Potential for breeding – members of the same species can breed together naturally and produce fertile offspring.

- Existing species are thought to have evolved from other species. One method of speciation involves the formation of a hybrid, the offspring from two different species, which then increases the number of sets of chromosomes in its cells (polyploidy). This method of speciation is thought to be responsible for the evolution of most species of flowering plants, but not of animals.

- Another method of speciation involves the reproductive isolation of two or more groups of a single species. Different pressures of natural selection on the isolated groups results in different frequencies of alleles in the groups. If these changes in allele frequency are great enough to prevent interbreeding, new species are formed.

- In the latter method of speciation, allopatric speciation occurs when the barrier to reproduction is geographical separation. Sympatric speciation occurs if the groups are reproductively isolated without geographical separation.

- Species are organised into biological categories in which the members show decreasing similarity to each other. Species are grouped into genera, which are then grouped into families. Families are grouped into orders, orders into classes, classes into phyla and phyla into kingdoms.

- The biological name of any species is a binomial, formed from the name of the genus and the name of the species. Through international convention, the name of the genus is written with a capital initial letter and the name of the species is written with a small initial letter. These names are printed in italics or, if hand-written, are underlined. The biological name for humans is *Homo sapiens*.

- There are five biological kingdoms: Prokaryotae, Protoctista, Fungi, Plantae and Animalia.

- The kingdom Prokaryotae contains all the bacteria. These organisms have prokaryotic cells (no true nucleus) and cell walls made of peptidoglycans. Some prokaryotes are heterotrophic and others are autotrophic, obtaining their nutrition by photosynthesis or by chemosynthesis.

- Organisms are classed in the kingdom Protoctista if they cannot be classed into one of the other four kingdoms. Protoctists have eukaryotic cells. They include organisms that resemble single-celled animals, organisms that resemble single-celled, filamentous or colonial plants and multicellular organisms, such as seaweeds.

- Fungi have eukaryotic cells, which are surrounded by cell walls made of chitin. They are heterotrophic, with most releasing enzymes onto their surroundings and absorbing the digested food materials into their cells. Most fungi are made of thread-like hyphae, which grow horizontally in the food source, but have erect hyphae, which carry their reproductive spores. The pigment in the spores produces the colour that we see on mouldy food. Yeast is a unicellular fungus that is used extensively in industry.

- Members of the phylum Plantae have eukaryotic cells surrounded by a cellulose cell wall. The phylum includes liverworts, mosses, ferns, conifers and flowering plants. The life cycle of plants is complex, containing two different types of adult. The haploid gametophyte adult reproduces sexually to form a zygote. The diploid adult that develops from a zygote is the sporophyte. It reproduces asexually to produce spores that germinate into gametophytes again.

● All members of the animal kingdom (Animalia) have eukaryotic cells that lack a cell wall. Animals always develop from a multicellular zygote that forms a hollow ball, called a blastula, at some stage. Humans belong to this kingdom.

# Assignment

### When is a primrose not a primrose?

In this chapter you have seen that it is not always easy to decide whether two organisms belong to the same or to different species. We will now consider this problem in a little more detail by looking at a plant which is common throughout much of Britain – the primrose. Some of the material in this assignment is about plant reproduction. Although this will probably be unfamiliar to you, you should be able to understand the principles involved from the information you have been given.

1 Figure 3.17 shows a clump of primroses. Use the information in this chapter to describe how you could show that the yellow-flowered plants and the pink-flowered plants belong to the same species.

*(3 marks)*

In many parts of Britain another species of *Primula* may be found growing near primroses. This is the cowslip. Figure 3.18 shows primroses, cowslips and the hybrids which sometimes form between these plants.

Even in areas where primroses and cowslips are common, hybrids between them are relatively rare. This may be the result of slightly different ecological requirements – cowslips are generally found in more open areas. It might also be due to cowslips generally flowering later than primroses.

In experimental conditions, primroses and cowslips readily produce hybrids but this only happens when the female parent is a cowslip and the pollen comes from a primrose. The reason for this is associated with the way in which flowering plants reproduce. When fertilisation

**Figure 3.17**
Primrose flowers are usually pale yellow but occasionally plants with pink flowers are found. Pink flowered primroses are quite common in some parts of Wales

**Figure 3.18**
(a) Primroses (the plants with larger paler flowers) are growing together with cowslips on this bank. (b) A hybrid plant growing nearby. It has some characteristics of its primrose parent and some characteristics of its cowslip parent

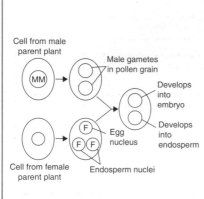

Cell from male parent plant

MM

Male gametes in pollen grain

Develops into embryo

Egg nucleus

Develops into endosperm

F

F  F

Cell from female parent plant

Endosperm nuclei

**Figure 3.19**

takes place in a flowering plant, the pollen grain provides two male gametes. One of these male gametes fuses with the egg nucleus to become the embryo. The other fuses with the two endosperm nuclei. The endosperm nucleus divides by mitosis and gives rise to endosperm. This is a tissue which provides a food source for the developing embryo.

2   Copy the diagram in Figure 3.19. Use the information in the paragraph above to complete this diagram to show the chromosomes in each of the nuclei. Use **M** to represent one set of chromosomes in or from the male parent and **F** to represent one set of chromosomes in or from the female parent. Some stages have been put in to help you.

*(4 marks)*

Now we will explain why we only get hybrids if the cowslip acts as the female parent. Primroses and cowslips have different rates of seed development. Primrose seeds develop more slowly than cowslip seeds. Not only does the whole seed develop more slowly but so does the embryo and the endosperm. Look at Table 3.6. This table shows the sets of cowslip chromosomes (**C**) and the sets of primrose chromosomes (**P**) in possible hybrids.

Table 3.6
The number of sets of cowslip chromosomes and primrose chromosomes in the endosperm and in the embryo will depend on whether the female parent is a cowslip or a primrose

| Female parent | Number of sets of cowslip chromosomes (C) and primrose chromosomes (P) in a nucleus from | |
| --- | --- | --- |
| | endosperm | embryo |
| Cowslip | CCP | CP |
| Primrose | CPP | CP |

3   (a)   The endosperm formed when the female parent is a primrose develops more slowly than the endosperm formed when the cowslip is the female. Use the information in the table to explain this difference.

*(2 marks)*

(b)   Experiments have shown that when the female parent of a primrose-cowslip cross is a primrose only about 1% of the seeds formed have embryos present. When the female parent of this cross is a cowslip, between 80 and 100% of the seeds contain embryos. Explain these findings.

*(2 marks)*

4   Do you think that primroses and cowslips are the same or different species? Use evidence from this assignment to support your answer.

*(4 marks)*

# Examination questions

1 The Hawaiian islands are a chain of volcanic islands in the middle of the Pacific Ocean. **Table 3.7** compares the number of species of some different groups of insect in Hawaii and in the British Isles.

| | Number of native species | |
| --- | --- | --- |
| | **Hawaii** | **British Isles** |
| Flies belonging to the genus *Drosophila* | 800 | 37 |
| Total number of species of fly | 800 | 5 950 |
| Total number of species of insect | 6 500 | 21 833 |

Table 3.7

Suggest an evolutionary explanation for each of the following observations.

(a) All the flies native to Hawaii belong to the same genus.

*(1 mark)*

(b) When the volcano erupted, lava poured out. This left isolated patches of vegetation surrounded by large areas of bare lava. There are slight differences between the populations of flies in neighbouring patches of vegetation even though they belong to the same species.

*(3 marks)*

(c) There are more species of *Drosophila* in Hawaii than there are in the British Isles.

*(2 marks)*

2 The diagram shows how four species of pig are classified.

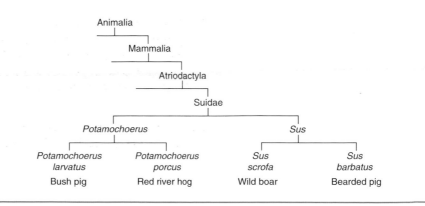

(a) (i) To which family does the red river hog belong?

*(1 mark)*

(ii) To which genus does the bearded pig belong?

*(1 mark)*

(b) Some biologists think bush pigs and red river hogs belong to the same species. The list below summarises some features of the biology of bush pigs and red river hogs.

- The bush pig has a body length of 100–175 cm and a mass of 45–150 kg. The red river hog has a body length of 100–145 cm and a mass of 45–115 kg.
- The red river hog is found in West Africa. The bush pig is found in East Africa.
- Both animals are omnivorous but feed mainly on a variety of underground roots and tubers.
- The ranges of these animals overlap in Uganda. In this area populations of animals which have all characteristics intermediate between those of bush pigs and red river hogs have existed for many years.

Do you think that bush pigs and red river hogs belong to the same or to different species? Explain how the information above supports your answer.

*(3 marks)*

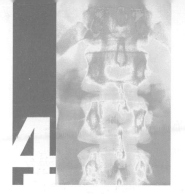

# Numbers and Diversity

Yanamamo is an area of tropical rain forest in the Peruvian Amazon. It probably has more species of animals and plants than anywhere else on Earth. A single hectare of the forest contains almost 300 different species of tree. Compare that with the United Kingdom where there are only 33 native tree species! These trees support huge numbers of animals. A biologist working in this area found a single tree, for example, which had 43 different species of ant living on it – that's about the same number as in the whole of the United Kingdom.

**Figure 4.1**
This tree has recently been felled. It has left a space where light can penetrate to the forest floor. Plants such as macaranga rapidly become established in areas like this

Not only are there enormous numbers of different species in a tropical rain forest but the relationships between them are often very complex. Macaranga is a fast-growing tree. It springs up wherever taller trees fall and allow light to reach the forest floor (Figure 4.1). Macaranga has hollow stems in which a particular species of ant lives. These ants 'farm' scale insects. The scale insects feed on the tree sap and produce a sugary liquid which is readily consumed by the ants. Macaranga trees with ants living in them are usually free of climbers and other plants living on them because the ants bite off any foreign shoot or leaf coming into contact with their host tree. In addition, experimental work has shown that once ants are removed, macarangas are attacked by many species of leaf-eating insect.

When humans clear patches of forest, species are lost and this may have far reaching effects. Jackfruit (Figure 4.2) is a tree which is pollinated by nectar-feeding bats. These bats need other species of tree on which to feed for those months of the year when jackfruit is not in flower. If you fell the forest, there will be no bats and no jackfruit.

**Figure 4.2**
Jackfruit, one of many tropical plants which are pollinated by bats. A good crop of jackfruit depends on a good population of bats

The huge number of different species and the complex biological relationships between them make tropical rain forests very difficult places to study. We find, however, that there are a number of basic principles which apply just as readily to a rain forest as they do to a seashore or a pond in Britain. In the next three chapters we shall look at these ecological principles in a little more detail.

An understanding of the ecology of a particular area – seashore, pond or rain forest – requires us first of all to gain some idea of the numbers of organisms found there and how they are distributed. Then we can begin to use our understanding of ecological principles to explain these observations. In this chapter we will start by looking at some of the different methods biologists use to find out about numbers and distribution. We will then consider two basic ecological ideas that help us to explain differences in distribution – diversity and succession.

## Organisms and their environment

Nettles are common plants. You can find them growing in woodland, by the side of ponds and streams and in areas of waste land in towns and cities. Look at Figure 4.3. It shows a clump of nettles and some of the animals which can be found living on them.

**Figure 4.3**
Some of the common organisms associated with a clump of nettles

Peacock butterfly (larva)

Peacock butterfly

Dark bush cricket

7-spot ladybird

7-spot ladybird (larva)

2-spot ladybird

Nettle aphid

The **environment** of the organisms in this clump of nettles is the set of conditions which surrounds them. It consists of an abiotic component, and a biotic component. **Abiotic factors** make up the non-living part of the environment. Nettles grow particularly well, for example, where there is a high concentration of phosphate in the soil. Warm, humid conditions result in large populations of aphids on the leaves. The concentration of mineral ions such as phosphate, temperature and humidity are all abiotic factors. **Biotic factors** are those relating to the living part of the environment. The number of aphids on these nettle plants will be affected by the number of ladybirds present since ladybirds feed on aphids. Predation by ladybirds is a biotic factor.

A **population** is a group of organisms belonging to the same species. The members of a population are able to breed with each other, so the two-spot ladybird population consists of all the two-spot ladybirds found in a particular area at the same time. The two-spot ladybirds found on our clump of nettles will form a different population from the ladybirds found on another clump two or three hundred metres away. A **community** is the term used to describe all the populations of different organisms living in the same place at the same time. When we talk about the community of organisms living in this nettle patch we mean the nettles, the aphids, the ladybirds and all the other organisms that we have not mentioned, such as the fungi and the bacteria which live in the soil round the roots of the nettles.

An **ecosystem** is made up of a community of living organisms and the abiotic factors which affect them. It is sometimes convenient to think of ecosystems as being separate from one another but they seldom are. A small bird for example, such as a blue tit, may feed on insects in a nettle patch but it might nest in a nearby wood and, in winter, visit bird tables on a housing estate.

**Q** 1 **Suggest how increased use of pesticides on farmland in North America could result in an increase in the concentration of pesticides in the tissues of penguins living in the Antarctic.**

There are two more terms we often use to explain ecological ideas. The **habitat** of an organism is the place where it lives. Its **niche** describes not only where it is found but what it does there. It is a description of how an organism fits into its environment. We can describe the niche of the two-spot ladybird, for example, in terms of the abiotic features of the habitat in which it lives, such as the temperature range it can tolerate and the position on the nettle plant where it is found. We can also bring into our description some idea of its feeding habits, referring to the size and species of aphids that it eats.

## Counting and estimating

It is not often that we can find the size of a population by counting all the organisms of a particular species. It can sometimes be done if the animal or plant is comparatively rare, very conspicuous or lives in a small area (Figure 4.4).

If it is not possible to count every single organism of a particular species, we need to take samples. Ecologists usually base their estimates of populations on samples. Whatever method is used, however, we must make sure that our samples are representative of the population as a whole. In order to be sure of this, they must be large enough and must be taken at random.

### Sample sizes

The larger the size of the sample the more reliable the results, but a very large sample may be too large to study accurately in the time available. Small samples, on the other hand, can be quite different from the rest of the population. We have to strike a balance. Look at the graph in Figure 4.5. It shows that a very small sample has very few species in it. As the sample size increases, the number of species it contains increases. There comes a time, however, when the number of species does not increase significantly, however much bigger the sample. This sample size is obviously representative of the population as a whole.

**Figure 4.4**
Mute swans are conspicuous birds and they are confined to aquatic habitats. It is possible to estimate the population of mute swans fairly accurately by counting all the birds in a particular area. There are likely to be errors, however, resulting from birds moving round or being present in areas where access is difficult

**Figure 4.5**
The number of species present in samples of different sizes. At the point marked **X**, the number of species present does not increase much more. This sample size is representative of the population as a whole. A larger sample size would simply mean more work

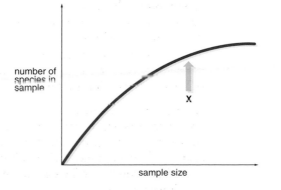

number of species in sample

sample size

**Figure 4.6**
A quadrat frame being used to sample organisms on the seashore. In this investigation, a random sample of limpets is being measured

## Random sampling

If we do not take samples at random, results will obviously be biased. Quadrat frames (Figure 4.6) are often used to sample vegetation or sedentary animals living on a seashore. If we use a quadrat frame for sampling, it is important that we use a method (Figure 4.7) that will result in it being placed genuinely at random in our study area.

**Q 2** **Which of the methods below would allow samples to be taken at random?**
**A Closing your eyes, turning on the spot and throwing a quadrat frame over your shoulder**
**B Picking numbers out of a hat to give coordinates on a grid**
**C Placing quadrat frames at 5 metre intervals**

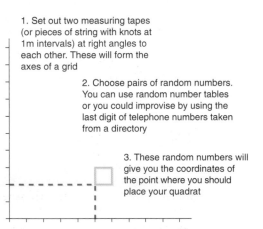

1. Set out two measuring tapes (or pieces of string with knots at 1m intervals) at right angles to each other. These will form the axes of a grid

2. Choose pairs of random numbers. You can use random number tables or you could improvise by using the last digit of telephone numbers taken from a directory

3. These random numbers will give you the coordinates of the point where you should place your quadrat

**Figure 4.7**
Placing quadrats at random. Where possible, a method like this should be used. Throwing the quadrat frame over your shoulder will not result in genuinely random sampling

## Quadrats and transects

Two important ways of sampling involve the use of quadrats and transects (Table 4.1). Both of these are generally used for plants but they can be used for other organisms which do not move about much – such as many of those that live on the seashore. There are three measures commonly used to describe the distribution of organisms once the quadrat or the transect is put in position. These are described in Figure 4.8.

(a) population density

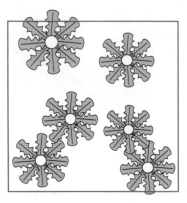

This quadrat measures 0.5m × 0.5m. It contains 6 dandelion plants. The population density of dandelions would be 24 plants per m². To get an accurate figure you would need to collect the results from a large number of quadrats.

**Figure 4.8**
Three measures commonly used to describe the distribution of plants and other organisms. These are (a) population density, (b) frequency and (c) percentage cover

(b) frequency

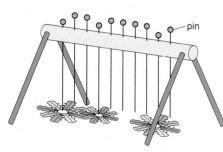

pin

This point quadrat frame is being used to measure frequency. The pins of the frame are lowered. Suppose 3 out of 10 pins hit a dandelion plant. The frequency of dandelion plants will be 3 out of 10 or 30%.

(c) percentage cover

Percentage cover measures the proportion of the ground in a quadrat occupied by a particular species. The percentage cover of the dandelions in this quadrat is approximately 30%.

| Method | What it is used for | How it is used |
|---|---|---|
| **Frame quadrat** <br> A sample area marked out in order to study the organisms it contains. | Usually used to study the distribution of plants in a fairly uniform area. | Placed at random and used to find population density, frequency or percentage cover. In some investigations, such as those involving succession or the effects of grazing, permanent quadrats may be used. They remain in place for many years. |
| **Transect** <br> A line through a study area along which samples are taken. | Used where the environment gradually changes and species vary. | Placed so that it follows the gradient, for example up a seashore or from full sunlight into dense shade. Quadrats may be used at regular intervals along the transect to sample in more detail. |

Table 4.1

Sampling methods – what for and how

## Mark–release–recapture

Most animals move around so it would not be possible to estimate their population size using a quadrat. We can use the mark–release–recapture method. This relies on capturing a number of animals and marking them so that they can be recognised again. They are then released. Some time later a second sample is trapped and the numbers of marked and unmarked animals recorded. From the data collected, the size of the population can be estimated. The calculation is very straightforward and relies on the fact that the proportion of marked animals in the second sample will be the same as the proportion of marked animals in the whole population (Extension box 1).

| Extension box 1 | **Using the mark–release–recapture method to estimate the size of a population of grasshoppers** |
| --- | --- |

Number of grasshoppers caught, marked and released      56

Number of marked grasshoppers in second sample      16

Total number of grasshoppers in second sample (marked + unmarked) 48

$$\text{Proportion of marked grasshoppers in sample} = \text{Proportion of marked grasshoppers in population}$$

$$\frac{\text{Number of marked grasshoppers in sample}}{\text{Total number of grasshoppers in sample}} = \frac{\text{Number of marked grasshoppers in population}}{\text{Total number of grasshoppers in population}}$$

$$\frac{16}{48} = \frac{56}{\text{Total population}}$$

$$\text{Total population} = \frac{48 \times 56}{16}$$

$$= 168$$

**Q** 3 A sample of 40 trout in a fish pond were netted and each fish was marked on one of the fins. They were then released back into the pond. One week later a second sample was netted. It contained 17 marked trout and 41 unmarked trout. Estimate the size of the trout population in the pond.

The mark–release–recapture method involves making a number of assumptions. These are:

- The number of animals in the population does not change between marking and releasing animals trapped in the first sample, and capturing the second sample. It will not work if lots of animals are born or hatch from eggs, or if large numbers die in the time between the two samples. It will not work either if the animal concerned is migrating and large numbers are either entering or leaving the study area.

- When the marked animals are released, they must mix thoroughly in the population.

- Marking must not affect the animal in any way. A large blob of white paint put on the back of a grasshopper, for example, will make it conspicuous and more likely to be eaten by a predator.

**Q** 4 **Blue tits breed during the spring. During the breeding season, pairs of birds establish territories which they defend from other blue tits. Give two reasons why the mark–release–recapture technique would not be suitable for estimating the population of blue tits in a wood in springtime.**

## Diversity

Biodiversity is a word frequently used by politicians and, like all words used by politicians, its meaning is not always clear! A biologist uses the word 'diversity' and means something very specific. Diversity is a way of describing a community in terms of the numbers of species and the number of organisms present. Different communities differ in their diversity. In the introduction to this chapter you read about some of the organisms present in a tropical rain forest. Tropical rain forests have a very high diversity. On the other hand, many Arctic communities (Figure 4.9) have a low diversity.

**Figure 4.9**
This area, high in the mountains of Norway, has a very low diversity of living organisms. It is characterised by a small number of species, although many of these are very numerous

As scientists, we try to describe diversity in numerical terms. In this way we can compare different communities. We will now look at how we can do this by calculating the **index of diversity, d**, from the formula:

$$d = \frac{N(N-1)}{\sum n(n-1)}$$

In this formula, **N** is the total number of organisms of all the species present in the community and **n** is the total number of organisms of each individual species. The Greek letter $\sum$ (sigma) means 'the sum of ...' so, in the bottom line of this formula we have to work out **n(n −1)** for each species and then add all these separate figures together. Extension box 2 shows a worked example.

**Extension box 2**

## Index of diversity – worked example

The figures in Table 4.2 show the birds visiting a bird table on a cold dry day and again, two days later, when it was raining heavily.

| Species | Maximum number of birds around bird table in one hour period | |
|---|---|---|
| | on dry day | on wet day |
| Blue tit | 8 | 1 |
| Coal tit | 1 | 1 |
| Great tit | 3 | 1 |
| Robin | 1 | 1 |
| House sparrow | 5 | 2 |
| Greenfinch | 3 | 1 |
| Starling | 7 | 7 |
| All species | 28 | 14 |
| Index of diversity, d | 5.8 | 4.1 |

**Table 4.2**
Number of birds visiting a bird table on a cold dry day and on a rainy day

We will look at the way in which we calculate the figures in the first column. The formula is:

$$d = \frac{N(N-1)}{\sum n(n-1)}$$

We can substitute the figures for N and n from the information in Table 4.2

$$d = \frac{28(28-1)}{8(8-1) + 1(1-1) + 3(3-1) + 1(1-1) + 5(5-1) + 3(3-1) + 7(7-1)}$$

$$d = \frac{28 \times 27}{(8 \times 7) + (1 \times 0) + (3 \times 2) + (1 \times 0) + (5 \times 4) + (3 \times 2) + (7 \times 6)}$$

$$d = \frac{756}{130} = 5.8$$

You might have thought 'Why do we not simply compare the number of species present? Surely, this would give us an idea about diversity and would save the need for a lot of tedious calculation.' The answer is that it does provide information in some cases, such as in comparing the number of species in a tropical rain forest with the number in an Arctic ecosystem. But, it does not take into account the fact that many species will be rare and only likely to be encountered in very small numbers. Look at Table 4.2 again. You have the same number of species in both columns. In the column representing the wet day, however, 5 out of the 7 species were encountered once only. An index of diversity gives us a much better idea of differences between communities in situations like this.

**Figure 4.10**
The seaweeds growing on the seashore and the animals living in them form distinct zones. There are fewer species living in the harsher conditions on the upper shore than in the less severe conditions on the lower shore

**Figure 4.11**
Abiotic conditions vary on this seashore. On the lower shore, organisms are covered by sea water for most of the time. On the upper shore, they are exposed to the air. Few marine species can tolerate these harsh conditions

On its own, the index of diversity tells us very little. It is important, however, in allowing us to make comparisons. In the example in Extension box 2, we can begin to see a pattern. With more observations, we might be able to say convincingly that the wetter the weather, the lower the diversity of birds visiting a garden bird table. By looking at a lot of different communities, ecologists have established a very important principle.

- The harsher the environment, the fewer the species present and the lower the diversity. In harsh environments it is generally abiotic factors which determine the species present.

- The less harsh the environment, the more species there are and the greater the diversity. In these environments it is often biotic factors such as competition which determines whether particular species can survive.

We will illustrate this principle by looking at the distribution of seaweeds on a rocky seashore such as that shown in Figure 4.10.

Seaweeds found on the upper shore are only covered in sea water for the small part of the day around high tide. For the rest of the day, they are exposed to the air. In these conditions the temperature undergoes considerable variation, evaporation of salt water and rain may bring about large changes in salt concentration, and long, hot summer days are likely to have a considerable drying effect. These are harsh conditions and very few seaweeds can survive them (Figure 4.11). The index of diversity for these seaweeds will be very low and it will be their tolerance of abiotic conditions which will determine the height up the shore to which they are found.

**Q 5** **What data would it be necessary to collect to calculate an index of diversity for seaweeds growing on the upper shore?**

Now we will look at the low-tide level. Seaweeds growing here will be covered in sea water for most of the day. Conditions are much more stable and not subject to large fluctuations in temperature and salt concentration. There is a much higher diversity of seaweeds. The distance they extend down the shore is likely to be due to biotic factors such as competition with other seaweeds. We will explore this principle further in Extension box 3 and in the last section in this chapter.

**Extension box 3**     **Diversity and freshwater pollution**

In this chapter we have been looking at an important ecological principle – the more extreme an environment, the fewer the species of organism found and the lower the diversity. We can apply this principle to pollution. Polluted habitats are extreme environments in which the diversity of organisms is low. In the assignment in Chapter 5 we will look at the way in which air pollution by sulphur dioxide affects the distribution of lichens. Here, we will consider how diversity can be used to study the amount of organic pollution in freshwater streams and rivers.

We can monitor organic pollution by measuring abiotic factors such as the concentration of dissolved oxygen in the water. With a suitable meter this can be done rapidly, but it only gives us a picture of the pattern of pollution at a particular time. If we want to get an overall picture, we need to take a lot of measurements, or we have to use another technique.

**Figure 4.12**
*Chironomus* is an insect whose larvae are very tolerant of organic pollution. The graph shows that there are very large numbers of *Chironomus* larvae immediately downstream of the sewage outlet

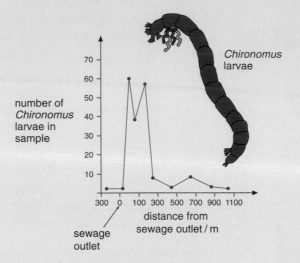

Organic pollution affects living organisms. We can use these organisms as indicators of the amount of pollution (Figure 4.12). The results obtained by looking at the distribution of a single species are not very reliable. In the example shown in Figure 4.12, there could be other factors which influence the distribution of *Chironomus* other than organic pollution. We get a better picture if we look at all the organisms present. There are various ways of doing this. The simplest is to collect a sample of living organisms and use it to calculate the sequential comparison index (SCI). This is given by the formula:

$$SCI = \frac{\text{number of sequences of organisms of the same species}}{\text{total number of organisms}}$$

Let us look at how it works. At an extremely polluted sampling point we may catch twenty individual organisms, all the same species. The SCI will therefore be 1/20 or 0.05 since there is only one sequence of organisms of the same species and twenty organisms in total. Suppose we sample at a less polluted point. We again catch a total of twenty organisms but they are of different species. We remove them from the net in the order:

A A A│B│A A A│B│C C C│D│E│F│A A A│B│F F│C C

Look at this list and you will see that there are twelve sequences of the same species – they are marked by vertical lines. The SCI will therefore be 12/20 or 0.6.

This method has a big advantage over other methods. It only requires a person to be able to recognise that organisms differ from each other. It does not rely on identification so it can be used by a non-biologist. The disadvantage is that it is rather crude and unreliable. In addition, it is influenced by other sorts of pollution. Because of this biologists have produced other measures called biotic indices which are more closely related to organic pollution, but they still rely on measuring the diversity of living organisms (Figure 4.13).

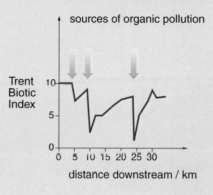

**Figure 4.13**
The Trent Biotic Index, an index based on the diversity of living organisms, at various points along the river North Esk in Scotland. The three main dips in the curve indicate sources of organic pollution

## Succession

Some of the organisms living in a community are gradually replaced by others. This is **succession**. It is an ecological process resulting from the activities of the organisms themselves. Over a period of time they modify their environment (Figure 4.14) and these modifications produce conditions better suited to the growth of other species.

In this book, we cannot describe all the examples of succession that you might encounter on a field course, so what we will do is to choose one particular example and use this to illustrate the principles which apply to them all. Sand dunes occur in many coastal areas and they show very clearly the process of succession. Obviously it is not possible to sit in one place and observe successional changes as they take place. The process is far too slow for that. What we can do, however, is to walk up from the sea shore into the sand dunes. As we do, we will pass through different areas which represent different stages in succession (Figure 4.15).

In understanding what has happened, it is useful to think again about the principles we considered when we discussed diversity. We will start by considering the mobile dunes. They get this name because they are continually changing shape as the wind scours the surface and blows the sand. This is a very harsh environment. Look at Table 4.3. It compares abiotic factors in these mobile dunes with those from the fixed dunes later in the succession.

**Figure 4.14**
Plants like bog bean grow in shallow water and trap mud around their roots. This eventually leads to drier conditions in which other species become established

| Abiotic factor | Mobile dunes | Fixed dunes |
|---|---|---|
| Mean wind velocity 5 cm above dune surface/km hour$^{-1}$ | 12.1 | 2.4 |
| Organic matter/% | 0.3 | 1.0 |
| Na$^+$/ppm | 8.5 | 4.2 |
| Ca$^{2+}$/ppm | 637.0 | 297.0 |
| NO$_3^-$/ppm | 48.0 | 380.0 |

**Table 4.3**
Abiotic factors in mobile and fixed dunes

**Figure 4.15**
Succession in sand dunes. As we walk up the shore we pass from an area of isolated plants growing in the drifting sand heaped round material left stranded by high tide, through an area dominated by marram grass to an area of fixed dunes with a complete covering of plants

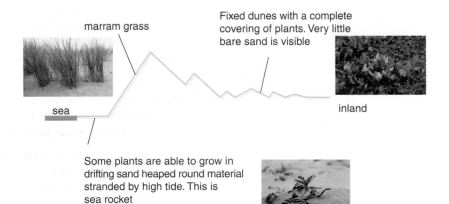

marram grass

Fixed dunes with a complete covering of plants. Very little bare sand is visible

sea

inland

Some plants are able to grow in drifting sand heaped round material stranded by high tide. This is sea rocket

The high wind speed has two important effects. It will pile sand on top of any plants growing there and it will lead to a high rate of water loss by transpiration (Chapter 11). There is very little organic matter, so the sand does not retain water very well. It will dry out rapidly after rain. Important soil nutrients such as nitrates are in short supply, while sea spray results in high concentrations of sodium and calcium ions. Very few plants can grow here, but one that does is marram grass, illustrated in Figure 4.15. Because it is one of the first plants to colonise the area, we refer to it as a **pioneer species**. Marram grass has a number of adaptations. Like many grasses, the leaves grow from an underground stem, but this stem does not grow horizontally, it grows vertically upwards. The more sand that is deposited on top, the more vigorously it grows.

**Q  6 How is the vertically growing underground stem of marram grass an adaptation for conditions found in mobile dunes?**

Marram grass also has many structural features that enable it to grow in dry conditions. You can find out more about these in Chapter 11.

We have here a good example of the principle we encountered when we discussed diversity.

- The harsher the environment, the fewer the species present and the lower the diversity. In harsh environments it is generally abiotic factors which determine the species present.

**Q 7 What is the value of the index of diversity in mobile sand dunes if the only organism present is marram grass?**

Now let us go inland to the area of fixed dunes. The first thing you notice is that there are many more species of plants growing here. Table 4.3 shows us that abiotic factors have changed and many of these changes are due to the activity of the living organisms growing on the mobile dunes. The roots and leaves of the grass form a windbreak. The wind velocity is lower so less sand is being blown around. Dead material falls from the marram grass and is broken down by the soil bacteria. The amount of humus is higher so the developing soil can retain moisture rather better and the concentrations of important nutrients such as nitrates rises. Rain is also beginning the process of leaching the soluble sodium and potassium ions from the surface layers of the soil. Other plants begin to appear and gradually replace marram grass as the dominant vegetation. We are beginning to get to a situation where

- The environment is less harsh, so there are more species and a greater diversity. In this environment it is biotic factors such as competition which determines whether particular species can survive.

The process of the organisms in the community affecting their environment in such a way that they are replaced by other species continues. The fixed dune community that we have been looking at is gradually colonised by bushes and then trees. Ultimately we reach a stage when no further change takes place. This is the **climax community** and, in Britain, it would generally be woodland of some sort.

We have described the process of succession as it takes place in sand dunes but there are other examples such as the edges of ponds and streams and bare rock where the process may be studied. The details may be different in each case but the principles are very similar. They are summarised in Figure 4.16.

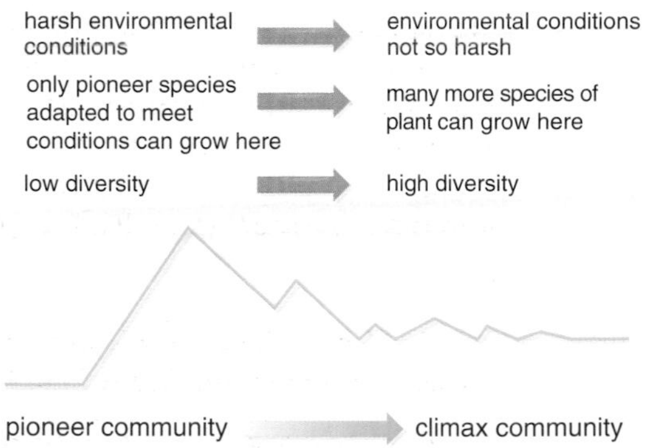

| harsh environmental conditions | → | environmental conditions not so harsh |
| only pioneer species adapted to meet conditions can grow here | → | many more species of plant can grow here |
| low diversity | → | high diversity |

pioneer community → climax community

**Figure 4.16**
A summary of the processes involved in succession

Succession does not always proceed all the way to a climax community however. It can be stopped by various factors such as human activity. In a series of experiments, control plots were compared with plots in which the plant cover had been partly destroyed by allowing motorcycles to ride over them. The results are shown in Figure 4.17.

**Figure 4.17**
Bar charts showing the effect of damage by motorcycles to the plants growing on sand dunes

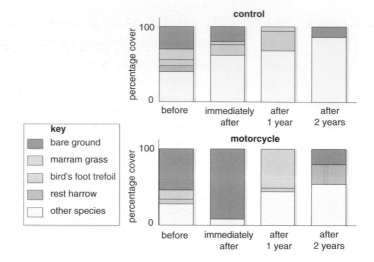

In the control plot, you can see that the process of succession has continued. Early colonisers such as marram grass and rest harrow have decreased in amount while the mean percentage of the plots covered with other species has increased. In the experimental plots, the initial effect of the motorcycles was to remove much of the plant cover. This allowed the spread of rest harrow.

Experiments like this show us that when we halt the process of succession, it rarely returns to where it began. In this example, it is quite easy to see why. The areas in which the vegetation has been partly removed by motorcycles are already quite different from those in which the marram grass became established. The sand contains more humus and is not so likely to dry out, it has a higher concentration of nitrate and lower concentrations of calcium and sodium ions, and there is less wind-blown sand. These are conditions in which rest harrow grows better than marram grass. Human activity, then, can alter the path of succession.

Another example where we see this happening is when we clear an area of tropical rain forest. If we leave this area, more forest grows in its place. This secondary forest, however, is quite different from the original forest. It tends to be denser and it contains different species.

Ecologists sometimes distinguish two types of succession. **Primary succession** occurs in places where previously there were no living organisms. The colonisation of bare sand dunes by marram grass and the changes which follow this are an example of primary succession. **Secondary succession** takes place in areas where there has already been a community of living organisms. An example of secondary succession is provided by the changes which follow when human activity damages the plant cover of sand dunes.

**Q** 8 Breeding elephant seals live on islands on which tussock grass grows. Their activity results in the death of this grass. When the seals leave at the end of the breeding season other plants grow in place of the tussock grass. Explain whether this is primary or secondary succession.

**Extension box 4**

## How old is a hedge?

Fields enclosed by hedges are a characteristic feature of the landscape in many parts of Britain. Some of these hedges are very old and may have been planted in Roman times. Some have been planted much more recently. How can we find the age of a particular hedge? We have seen that, during the process of succession, many changes take place in a community. Is there any way we can use our knowledge of these changes to suggest when a hedge was originally planted?

Look at Figure 4.18. This is a scatter diagram showing the number of different species of shrubs in random thirty metre lengths of hedge. The hedges in this study were all dated from historical records, so we have been able to plot the number of species against the exact age of the hedge

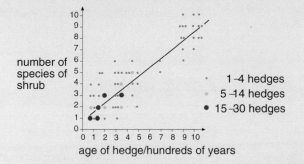

**Figure 4.18**
This scatter diagram shows a positive correlation between the age of a hedge and the number of species of shrubs it contains

You can see that there is an obvious trend. The older the hedge, the more species it contains. In approximate terms, there is one additional species every hundred years. So, a hedge containing only hawthorn is likely to be less than 100 years old; a hedge with three different species may have been planted at the beginning of the nineteenth century; and a hedge with six species probably dates back to somewhere around 1500.

Of course we have to be careful how we use information like this. The scatter diagram also shows us that there is a lot of variation in the number of species of shrubs in hedges of the same age. There are many reasons for this variation. More than one species might have been planted in the first place; variation in the soil and climatic conditions may have a considerable effect on the species which become established during the process of succession; and different hedges are different distances away from woods and other sources of seeds. Because of all these reasons, the best we can probably do is suggest a date but to remember that for a particular hedge we could be as much as 200 years out.

**Q** 9 How would you expect the diversity of shrubs in an old hedge to differ from that in a recently planted hedge?

# Summary

- Random sampling with quadrats and counting along transects may be used to obtain quantitative data about the distribution of organisms.

- The number of organisms that move around can be estimated with the mark–release–recapture technique.

- Diversity is a measure of the number of species and number of individuals in a community.

- The harsher the environment, the fewer the species present and the lower the diversity. In harsh environments it is generally abiotic factors which determine the species present.

- The less harsh the environment, the more species there are and the greater the diversity. In these environments it is often biotic factors such as competition which determine whether particular species can survive.

- Succession involves colonisation by pioneer species. Subsequent changes lead to the establishment of a climax community.

# Assignment

### How many tigers are there?

The tiger is one of the most impressive of all wild mammals. Unfortunately, it is also one of the rarest. Poaching for skins and for body parts used in traditional Chinese medicine, and the inevitable conflict with ever-increasing human populations, have led to a dramatic decline in wild tigers. Whether the tiger will survive as a wild mammal long into this century is open to considerable doubt.

In trying to conserve animals such as tigers, one of the first things biologists must do is to monitor the population as accurately as possible. In this assignment we will look at a study carried out in the Way Kambas National Park in Sumatra (Figure 4.19).

1  Apart from poaching, suggest why tigers are particularly threatened in this national park.

*(4 marks)*

2  The map shows that this national park is a mixture of dense secondary forest and tall grass. Suggest why it would be difficult to produce an accurate count of the tigers living in the park.

*(2 marks)*

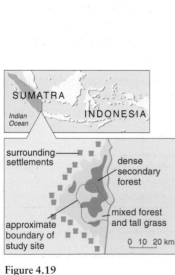

**Figure 4.19**
A map of the Way Kambas National Park on the Indonesian island of Sumatra

An ecological study was started in August 1995 using camera 'traps'. Automatic cameras were set up at selected points along trails used by tigers in the study area. These cameras took a photograph each time an animal passed through an infra-red beam crossing the trail. The cameras had built-in flash and were spread evenly through the study area.

3 Why was it necessary for the cameras to have built-in flash in order to get accurate results ?

*(1 mark)*

The pattern of stripes on a tiger is unique. By analysing the photographs taken by the cameras, it proved possible to identify individual animals. The graph in Figure 4.20 shows some of the results obtained from this study. Curve **A** shows the cumulative number of tigers photographed. Curve **B** shows the cumulative number of tigers photographed which were thought to be resident in the study area.

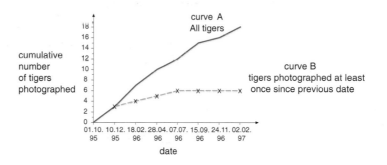

Figure 4.20

Tigers are territorial animals. They spend most of their time in a relatively small area which they defend against other tigers. The territory of a male tiger is considerably larger than the territory of a female tiger.

4 How many tigers had territories in the study area? Explain the evidence from the graph which supports your answer.

*(3 marks)*

Table 4.4 below shows some more data from this investigation.

| Tiger | Number of times photographed (P) | Number of camera locations at which animal photographed (L) | $\dfrac{P}{L}$ |
|---|---|---|---|
| A | 13 | 2 | |
| B | 43 | 11 | |
| C | 15 | 4 | |
| D | 23 | 8 | |
| E | 13 | 5 | |
| F | 19 | 11 | |

5 Copy and complete the table by calculating the ratio of the number of times a tiger was photographed to the number of camera locations at which it was photographed.

*(1 mark)*

6 (a) What does this ratio tell us about the size of a particular tiger's territory?

*(1 mark)*

(b) Two of the animals in the table were adult males. Use the information in this assignment to suggest which two.

*(1 mark)*

7 The study area was approximately 12% of the total park area. A suggestion was made that the total number of tigers in the park could be calculated from the following formula:

$$\text{Total number of tigers} = \frac{\text{Number of tigers with territories in the study area} \times 100}{12}$$

What assumptions does the use of this formula make? Suggest why using the formula would give an inaccurate estimate of the number of tigers in the park.

*(4 marks)*

# Examination questions

1 (a) Suggest **two** ways in which modern farming might affect the diversity of living organisms.

*(2 marks)*

(b) "Set-aside" is the common name given to a European policy under which farmers receive a subsidy for land taken out of cultivation. A study was carried out to investigate how the amount of time a set-aside field was left uncultivated would affect the species of birds feeding there.

Table 1

| Species | Number of birds of that species feeding in the field |
|---|---|
| Greenfinch | 12 |
| Goldfinch | 8 |
| Wood pigeon | 3 |
| Pheasant | 1 |

Table 1 on page 90 shows the number of birds of different species feeding in one field which had been left uncultivated for a year.

(i) Use the formula $d = \dfrac{N(N-1)}{\sum n(n-1)}$

where    d = index of diversity
           N = total number of organisms of all species
and      n = total number of organisms of a particular species

to calculate the index of diversity for the birds feeding in the field. Show your working.

*(2 marks)*

(ii) Explain why it is more useful in a study of this sort to record diversity rather than the number of species present.

*(2 marks)*

(c) **Figure 1** is a graph showing the relationship between bird species diversity and plant species diversity in this study.
**Figure 2** is a graph showing the relationship between bird species diversity and plant structural diversity for the same study.
Structural diversity refers to the different forms of plants such as herbs, shrubs and trees.

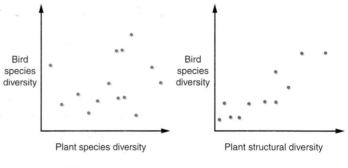

Figure 1             Figure 2

(i) Explain briefly how you could obtain the data that would enable you to calculate the diversity index for the species of plants growing on a set-aside field.

*(3 marks)*

(ii) Describe the difference in the relationships shown in **Figure 1** and **Figure 2**.

*(2 marks)*

(iii) Suggest an explanation for the relationship between bird species diversity and plant structural diversity and plant structural diversity shown in **Figure 2**.

*(2 marks)*

(d) In another study of fields taken out of cultivation, the figures shown in **Table 2** were obtained.

| Value of index of diversity for bird species | Time in years since cultivaton stopped |
|:---:|:---:|
| 2.1 | 5 |
| 3.2 | 15 |
| 5.6 | 20 |
| 4.1 | 25 |
| 4.8 | 40 |
| 9.4 | 60 |

Table 2

(i) Plot these data as a suitable graph.

*(4 marks)*

(ii) Predict what might happen to the bird species diversity in the study summarised in **Table 2** over the next 100 years. Explain how you arrived at your answer.

*(3 marks)*

2 The diagram shows a number of stages in an ecological succession in a lake.

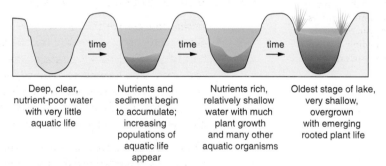

Deep, clear, nutrient-poor water with very little aquatic life
→ time →
Nutrients and sediment begin to accumulate; increasing populations of aquatic life appear
→ time →
Nutrients rich, relatively shallow water with much plant growth and many other aquatic organisms
→ time →
Oldest stage of lake, very shallow, overgrown with emerging rooted plant life

(a) Use information in this diagram to help explain what is meant by an ecological succession.

*(2 marks)*

(b) Give two general features which this succession has in common with other ecological successions.

*(2 marks)*

(c) A number of small rivers normally flow into this lake. These rivers flow through forested areas. Explain how deforestation of the area might affect the process of succession in the lake.

*(2 marks)*

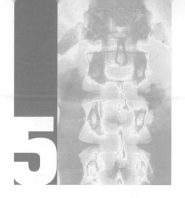

# Energy Transfer

The bottom of the Pacific Ocean is an inhospitable place. It is pitch dark; no light ever penetrates to its depths. And it is cold; the water remains all year round just above freezing. In a few areas, volcanic vents bubble out a mixture of sulphur-rich gases. The ocean floor is a place where you might think life could not possibly exist. But you would be wrong.

Around volcanic vents, bacteria are found. They make use of the rich source of sulphur-containing substances bubbling from the vents. They obtain energy from chemical reactions involving these substances and use it to build up the complex organic molecules that make up their cells. Various single-celled organisms feed on the bacteria and these, in turn, support a community of large worms and other invertebrate animals.

Far from being lifeless, the ocean depths contain many living organisms. These organisms depend on energy transferred from chemical reactions involving substances such as the sulphur-containing compounds emerging from volcanic vents. Or they may depend on energy contained in large organic molecules, which form part of dead organisms falling to the ocean floor (Figure 5.1).

This energy is then used to build up small molecules into the larger ones that make up cells and tissues. Communities of living organisms exist because chemical potential energy in large molecules is transferred from one organism to another along food chains and through food webs.

Figure 5.1
This deep-sea angler fish lives at great depths. It is part of a community of organisms that is able to survive because of a constant rain of organic matter drifting down from the sunlit water at the surface. Breakdown of this organic matter provides the energy necessary to sustain a whole community of organisms

In this chapter we shall look more closely at energy transfer. As we will not be looking at ocean depths, we shall be concerned mainly with another energy source, sunlight. As you can see from Figure 5.2, plants and other chlorophyll-containing organisms photosynthesise, using carbon dioxide and water to produce carbohydrates. In this process energy from sunlight is transferred to chemical potential energy in the glucose and other carbohydrates that are formed.

These carbohydrates can be used directly by the plants or they can be transferred to other organisms which either feed on plant material or break down dead plant tissue. They can also be used for respiration. During respiration, chemical potential energy in glucose is transferred to another molecule, ATP. ATP provides an immediate source of energy for various forms of biological work such as movement, the synthesis of large organic molecules and active transport.

Finally we need to remember that none of these processes is totally efficient. During respiration, for example, we cannot transfer all the chemical potential energy from a molecule of glucose into molecules of ATP. Some of the energy in the glucose molecule will inevitably be lost as heat. This is obviously important when we come to look at the transfer of energy from one organism to another along a food chain or through a food web.

**Q** 1 Some animals are cannibals and eat their own young. Use your knowledge of energy transfer to explain why there are no animals that only feed on their own young.

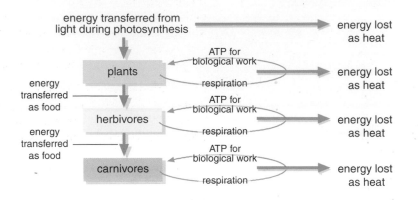

**Figure 5.2**
A summary of the way in which energy is transferred within and between organisms. Note that every time a transfer occurs, some energy is lost as heat

In this chapter we shall look at four main topics relating to this general theme of energy transfer.

● ATP as an energy source

● Photosynthesis

● Respiration

● Energy transfer and ecosystems

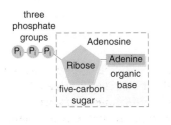

**Figure 5.3**
This simplified diagram of an ATP molecule shows the arrangement of its main components

## ATP as an energy source

Figure 5.3 shows a simplified diagram of an ATP molecule. You will see it has several components. **Ribose**, a five-carbon sugar, and **adenine**, an organic base, combine to give **adenosine**. Adenosine is joined in turn to three **phosphate groups**. When a single phosphate group is joined we have a molecule of adenosine monophosphate (AMP); two phosphate groups give us **adenosine diphosphate** (ADP); and three phosphate groups produce **adenosine triphosphate** (ATP).

In a living cell, ATP is normally produced by adding a third phosphate group to a molecule of ADP. This reaction requires a considerable amount of energy and this comes from chemical reactions such as those that occur in respiration and photosynthesis.

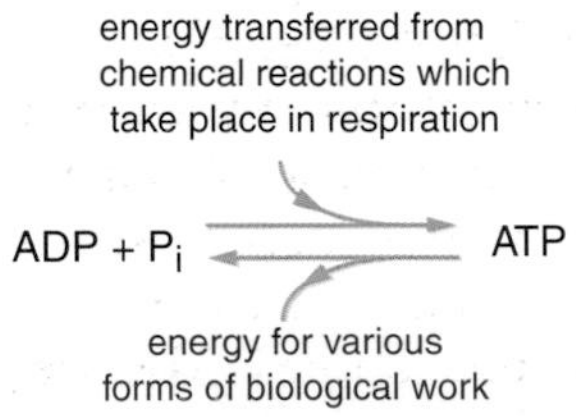

ATP breaks down to form ADP and a phosphate ion (written as $P_i$ for inorganic phosphate). This reaction involves hydrolysis of the ATP molecule and it releases a large amount of energy which can be used for various energy-requiring reactions. Since more energy is released than is used, some heat is always lost. ATP is a very convenient short-term store of chemical potential energy in living organisms. We can summarise the reactions involved in its formation and breakdown in the simple equation in Figure 5.4.

**Figure 5.4**
The formation and breakdown of ATP allows energy released in reactions such as those which take place in respiration to be made available for processes such as the synthesis of large molecules and active transport

Why do we need to transfer chemical potential energy from glucose molecules to ATP molecules? ATP is more useful than glucose as an immediate source of energy because:

● the breakdown of ATP to ADP and phosphate is a single reaction. It makes energy instantly available. The breakdown of glucose to carbon dioxide and water is, as we shall see when we look at respiration in more detail, a complex process involving many stages. It would take much longer to make energy available from glucose than from ATP.

● the breakdown of a molecule of ATP releases a small amount of energy, ideal for fuelling the energy requiring reactions which take place in the body. The breakdown of a molecule of glucose would produce more energy than is required.

**Q** 2 **It is sometimes suggested that a molecule of ATP can move around a cell more easily than a molecule of glucose because it is smaller. True or false?**

## Photosynthesis

### What is it all about?

Humans must have a supply of complex organic molecules to provide the chemical potential energy they need to build up new cells and tissues. These organic molecules come from our food – sometimes from other animals but ultimately from plants. It is only by photosynthesis that light energy can be converted into chemical potential energy and simple inorganic molecules such as carbon dioxide and water can be built up into organic ones. Ultimately we all depend on photosynthesis.

Photosynthesis is a complex process involving a number of separate reactions. It is useful to get an idea of the overall process before we look at any detail. There are two basic steps (Figure 5.5). In the **light-dependent reactions**, two substances are produced. Light energy is captured by chlorophyll and is transferred to chemical potential energy in ATP. The second substance, which we will look at in more detail in the next part of this chapter, is reduced NADP. In order to produce these substances, a molecule of water is split and oxygen is given off as a waste product. In the **light-independent reactions** ATP and reduced NADP are used in the conversion of carbon dioxide to carbohydrate.

**Figure 5.5**
The main steps in photosynthesis. The substances entering and leaving the main box can be arranged to give you the basic equation for photosynthesis

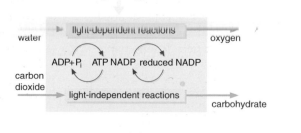

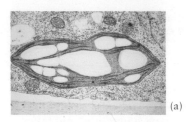

(a)

**Figure 5.6**
(a) A transmission electron micrograph of a chloroplast. This chloroplast is approximately 5 μm in diameter. (b) The chloroplast has a three-dimensional structure.

(b)

A simple model of the way in which the membranes inside a chloroplast are arranged. You will need some coins – 15 to 20 2p pieces will do fine and about 5 pieces of paper cut out like this:

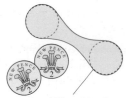

outline of 2p piece

Now arrange your coins in stacks. Vary the number of coins in a stack. Link them to each other with the pieces of paper you have cut out, like this:

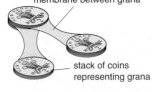

paper linking coin stacks. This represents the membrane between grana

stack of coins representing grana

# Chloroplasts, chlorophyll and light energy

You should remember from your AS course that chloroplasts are the site of photosynthesis. Each chloroplast (Figure 5.6) is surrounded by two plasma membranes. A system of membranes is also found inside the organelle. These membranes form a series of flattened sacs called **thylakoids**. In some places the **thylakoids** are arranged in stacks called **grana**. More membranes join the grana to one another.

The membranes which form the grana provide a very large surface for chlorophyll and other light-absorbing pigments. These pigments form clusters called photosystems (Figure 5.7). Each photosystem contains a chlorophyll molecule called a **primary pigment molecule**. Situated around this are several hundred **accessory pigment molecules**. These accessory pigment molecules include various different forms of chlorophyll and other light-absorbing pigments. The whole photosystem acts as a light-harvesting system. Light energy is captured and passed from one accessory pigment molecule to another before it finally reaches the primary pigment molecule.

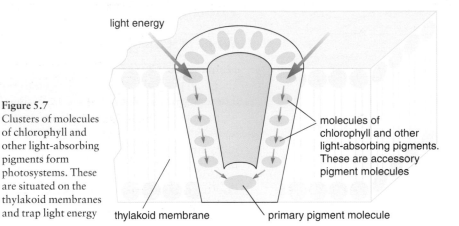

**Figure 5.7**
Clusters of molecules of chlorophyll and other light-absorbing pigments form photosystems. These are situated on the thylakoid membranes and trap light energy

light energy

molecules of chlorophyll and other light-absorbing pigments. These are accessory pigment molecules

thylakoid membrane

primary pigment molecule

## The light-dependent reactions

If we make a solution of chlorophyll and shine a bright light on it, it fluoresces. Instead of appearing green, it looks red in colour. What happens is that light energy raises the energy level of some of the electrons in the chlorophyll. These electrons leave the chlorophyll molecule. They lose most of this energy as light of a slightly different wavelength as they fall back into their place in the chlorophyll molecules. In a chloroplast, however, the electrons do not return to the chlorophyll molecule from which they came. They pass down a series of electron carriers, losing energy as they go. This energy is used to produce ATP and reduced NADP.

A whole series of reactions is involved which we can represent as a diagram (Figure 5.8). We will look at these reactions in a little more detail starting with light striking photosystem II.

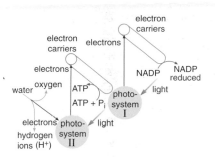

**Figure 5.8**
A summary of the light-dependent reactions of photosynthesis. The bullet points in the text should help you to understand this diagram

- Light strikes photosystem II. The energy levels of two of the electrons in the chlorophyll molecule which acts as the primary pigment molecule are raised. These electrons leave the chlorophyll molecule and pass to an electron acceptor. This results in the chlorophyll molecule now being positively charged.

- Photosystem II contains an enzyme which catalyses the breakdown of water in a process called **photolysis**.

$$H_2O \rightarrow 2H^+ + 2e^- + \tfrac{1}{2}O_2$$

The oxygen is given off as a waste product and the electrons replace those lost from photosystem II. We will look at what happens to the hydrogen ions later.

- The electrons which have been passed to the electron acceptor now pass down a series of electron carriers losing energy as they go. This energy is used to produce ATP.

- Light also strikes photosystem I and a similar sequence of events occurs to that which took place when light struck photosystem II. Electrons again leave the chlorophyll molecule which acts as the primary pigment molecule. They pass to an electron acceptor, leaving the chlorophyll molecule positively charged.

- The electrons which have passed down the electron carriers from photosystem II replace those lost from photosystem I.

- The electrons from photosystem I also pass down a series of electron carriers. They are used, together with the hydrogen ions from the photolysis of water to produce reduced NADP.

$$NADP + 2H^+ + 2e^- \rightarrow \text{reduced NADP}$$

**Q** 3 What happens to the electrons which come from:
(a) the water molecule
(b) photosystem I
(c) photosystem II?

## The light-independent reactions

In the light-independent reactions, ATP and reduced NADP are used to convert carbon dioxide into carbohydrate. These reactions get their name because they are independent of light. However, they cannot take place without products produced with light energy. Because of this, the light-independent reactions cannot continue for long in the dark.

**Q** 4 The light-independent reactions of photosynthesis are sometimes called the 'dark reactions'. Explain why this term is misleading.

In a cycle of reactions, carbon dioxide combines with a five-carbon sugar called **ribulose bisphosphate** (RuBP). Two molecules of a three-carbon compound, glycerate 3-phosphate (GP) are formed (Figure 5.9). The next step in the cycle is the conversion of glycerate 3-phosphate to triose phosphate. This reaction involves reduction and requires ATP and reduced NADP. Triose phosphate is a carbohydrate. Some of it is built

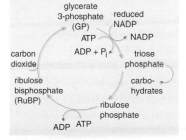

**Figure 5.9**
The light-independent reactions of photosynthesis form a cycle of reactions called the Calvin cycle. The diagram summarises the main steps in this cycle

energy in light reflected    energy from sunlight    energy lost as heat

energy in light passing through leaves and not striking chlorophyll

**Figure 5.10**
A potato crop like that shown here is able to convert around 2% of the energy in visible light into chemical potential energy in organic molecules

up into other carbohydrates, such as glucose and starch, and into amino acids and lipids. The rest of it is used to make more ribulose bisphosphate so the cycle can continue.

**Q** 5 ATP supplies energy for the conversion of GP to triose phosphate. Use Figure 5.9 to describe one other function of ATP in the light-independent reactions of photosynthesis.

## How efficient is photosynthesis?

A lot of light energy falls on the Earth's surface. Only a small part of this is actually use by plants and converted into chemical potential energy. Look at the potato crop shown in Figure 5.10. About half of all the visible-light energy falling on the plants in this crop is absorbed and converted into heat energy. Much of this heat energy evaporates water from the surface of the soil and from plant leaves during transpiration.

Of the remaining energy, just over 15% is reflected; and 32%, almost a third, is transmitted. It passes directly through the plant without striking any chlorophyll molecules on the way. As you can see, only a very small proportion of the incoming light energy ends up as chemical potential energy in substances produced by photosynthesis. For a crop grown in the UK, this is about 2%. We should also remember that not all of this energy goes into substances which form new tissues and cells. Quite a lot is used for respiration.

**Q** 6 In mid-summer the amount of light energy falling on an acre of farmland in the United Kingdom is approximately 18 800 kJ m$^{-2}$ day$^{-1}$. Use information in the paragraph above to calculate the amount of energy converted to chemical potential energy in the tissues of the crop.

---

**Extension box 1**

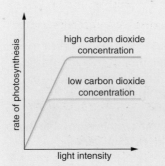

**Figure 5.11**
The effect of light intensity and carbon dioxide concentration on the rate of photosynthesis

## Any advance on 2%?

You may not be very impressed with the fact that a typical British crop plant only converts about 2% of the light energy falling on it into chemical potential energy. The figure of 2% though, is very much an approximation. Some organisms, such as the lichens which are described in the assignment at the end of this chapter are even less efficient than this; others, like the tropical crop, sugar cane, are rather better. In order to understand why the figures vary, we need to look at the graph in Figure 5.11. It shows us that at low light intensities the rate of photosynthesis is limited by light intensity. The evidence for this is that as you increase the light intensity, the rate of photosynthesis increases. At high light intensities, however, it does not make any difference if you increase the light intensity further. The rate of photosynthesis does not get any faster. At high light intensities it is something else, such as carbon dioxide concentration, which is limiting.

A crop of sugar cane (Figure 5.12) is able to convert 7 to 8% of the light energy it receives into chemical potential energy. It is probably the most productive of all plants. Why is it so efficient?

**Figure 5.12**
Sugar cane is a crop plant and like all crop plants it has been selected over many generations for particular characteristics. One of these is the efficiency with which it converts light energy into chemical potential energy in useful substances

- Light intensity is generally much higher in the tropics than in temperate regions. During the daylight hours, it is unlikely to limit the rate of photosynthesis. Tropical temperatures are also higher than those in temperate regions. The enzyme-controlled reactions involved with synthesis of organic molecules and growth are therefore likely to be faster in a tropical crop.

- Sugar cane is a grass. It therefore has tall upright leaves. This arrangement is very efficient at intercepting light; much more efficient than the horizontal arrangement found in many other plants.

- It is a cultivated plant grown in plantations. It is frequently irrigated and it is provided with fertiliser, so a lack of water and mineral ions are unlikely to limit its growth. In addition, weeds are controlled so the light energy goes to produce chemical potential energy in the sugar cane plants, not in other plants.

- You may remember from your AS course that tropical plants such as maize and sugar cane have a particular type of photosynthesis that is based on a special biochemical pathway, the C4 pathway. This pathway involves a series of biochemical reactions in which carbon dioxide is concentrated. This makes it much more efficient in hot, sunny conditions such as are found where sugar cane grows.

## Respiration

### What's it all about?

Respiration takes place in all living cells. It is a biochemical process in which organic substances are used as fuel. They are broken down in a series of stages and the chemical potential energy they contain is transferred to another molecule, ATP. This provides an immediate source of energy for various forms of biological work such as movement, the synthesis of large organic molecules and active transport.

You may be familiar with the simple equation that we sometimes use to summarise the process of respiration.

$$C_6H_{12}O_6 + 6O_2 \rightarrow 6CO_2 + 6H_2O + \text{Energy}$$

It is really an equation for the complete oxidation of glucose and is misleading in a number of ways. It shows the fuel as glucose. In many living cells, although the main fuel is glucose, fatty acids, glycerol and amino acids are also respiratory substrates and can be used for respiration. The equation also shows that oxygen is required. Respiration can, however, take place anaerobically, that is without the presence of free oxygen. Finally the equation fails to show that respiration involves a number of reactions in which the respiratory substrate is broken down in a series of steps, releasing a small amount of energy each time.

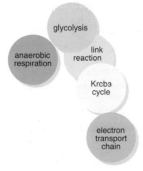

**Figure 5.13**
The breakdown of glucose in respiration. Aerobic respiration takes place in the presence of oxygen. Respiration can continue when there is no oxygen present. This is anaerobic respiration

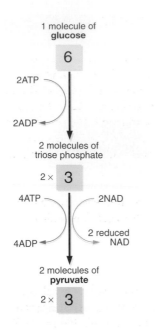

**Figure 5.14**
A summary of glycolysis. In this first step in respiration, a molecule of glucose is broken down to give two pyruvate groups. The boxes show the number of carbon atoms in some of the molecules and ions

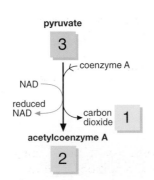

**Figure 5.15**
The link reaction. The boxes show the number of carbon atoms which are directly involved in the respiratory pathway. Acetylcoenzyme A contains more than two carbon atoms but only these two enter the next stage of respiration, the Krebs cycle

We will look in more detail at glucose breakdown. It involves the four steps shown in Figure 5.13.

## Glycolysis

Glycolysis (Figure 5.14) is the first step in the biochemical pathway of respiration. A molecule of glucose is broken down into two pyruvate groups, each of which contains three carbon atoms.

Glucose contains a lot of chemical potential energy but it is not a very reactive substance. In order to release this energy, we need to use some energy from ATP. In the first stage of glycolysis, a molecule of glucose is converted into two molecules of triose phosphate. This requires two molecules of ATP.

Triose phosphate is then converted to pyruvate. This process produces four molecules of ATP, two for each triose phosphate. The conversion of triose phosphate to pyruvate is an oxidation and involves the transfer of hydrogen to a carrier molecule. This carrier molecule or **coenzyme** is known as NAD. Adding hydrogen reduces NAD to reduced NAD.

Summarising glycolysis, for each glucose molecule we get two pyruvate groups, a net gain of two molecules of ATP and two molecules of reduced NAD. We will look at reduced NAD again in the section on the electron transport chain and see how it is used to produce more ATP.

## The link reaction

Pyruvate still contains a lot of chemical potential energy. When oxygen is available, this energy can be released in a cycle of reactions known as the **Krebs cycle**. The link reaction (Figure 5.15) is a term used to describe the reaction linking glycolysis and the Krebs cycle.

Pyruvate combines with coenzyme A to produce acetylcoenzyme A. This involves the loss of a molecule of carbon dioxide. It is also an oxidation reaction and results in the formation of another molecule of reduced NAD.

**Q** 7 **Complete the following equation which summarises the link reaction**

pyruvate + coenzyme A + NAD →

## The Krebs cycle

The Krebs cycle (Figure 5.16) is a cycle of reactions which involves a number of intermediate steps. We will only concern ourselves here with its main features.

- Acetylcoenzyme A produced in the link reaction is fed into the cycle. It combines with a 4-carbon compound (oxaloacetate) to produce a 6-carbon compound (citrate).

- In a series of reactions, the citrate is converted back to oxaloacetate. This involves the production of two molecules of carbon dioxide which is given off as a waste gas.

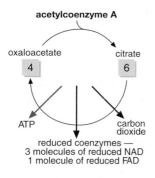

acetylcoenzyme A

oxaloacetate → citrate
4      6

ATP     carbon dioxide

reduced coenzymes —
3 molecules of reduced NAD
1 molecule of reduced FAD

**Figure 5.16**
The Krebs cycle plays a very important part in respiration. It is the main source of the reduced coenzymes which are used to produce ATP in the electron transport chain

- For each complete turn of the Krebs cycle, one molecule of ATP is produced.

- The reactions which form the Krebs cycle again involve oxidation. The hydrogens which are given off are used to reduce coenzymes. Each turn of the cycle produces three molecules of reduced NAD and one molecule of another reduced coenzyme, reduced FAD. The most important function of the Krebs cycle in respiration is the production of reduced coenzymes. These are passed to the electron transport chain where the chemical potential energy these molecules contain is used to produce ATP.

**Q**   8   **One molecule of glucose produces two pyruvate groups. How many molecules of reduced NAD are produced for each molecule of glucose respired?**

## The electron transport chain

The reduced NAD and reduced FAD produced as a result of oxidation now pass to chains of molecules situated on the internal membranes in the mitochondria. These are the electron transport chains (Figure 5.17). The hydrogen is removed from the two reduced coenzymes and is split into hydrogen ions ($H^+$) and electrons.

The electrons pass from one molecule to the next along the electron transport chain. At each transfer, a small amount of energy is released. This is used to pump the hydrogen ions out through the inner mitochondrial membrane into the space between the inner and the outer membrane. When the hydrogen ions return into the matrix of the mitochondrion they release their potential energy. This is used to produce ATP. The last molecule in the chain is oxygen. Oxygen combines with hydrogen ions and electrons to produce water.

**Figure 5.17**
As electrons pass down the electron transport chain, energy is released. Some of this energy is lost as heat but a lot of it goes to produce ATP. Each molecule of NAD entering the chain produces three molecules of ATP. A molecule of reduced FAD produces two molecules of ATP

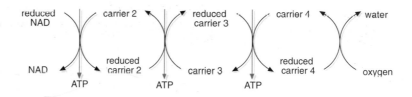

reduced NAD    carrier 2    reduced carrier 3    carrier 4    water

NAD    reduced carrier 2    carrier 3    reduced carrier 4    oxygen

ATP     ATP     ATP

## Anaerobic respiration

Sometimes there is not enough oxygen for an organism to respire using the pathway we have just described. Under these conditions it has to produce ATP by respiring anaerobically. The only stage in the anaerobic pathway which produces ATP is glycolysis, so it is not as efficient as aerobic respiration. The chemical potential energy from a single molecule of glucose can be used to produce 38 molecules of ATP in aerobic respiration. It can only produce 2 molecules of ATP in anaerobic respiration.

Look back at Figure 5.14. You will see that during glycolysis NAD is reduced. Reduced NAD is normally converted back to NAD when its hydrogen is transferred in the electron transport chain. This will only happen when oxygen is present. Obviously, if all the NAD in a cell was converted to reduced NAD, the process of respiration would stop. In anaerobic respiration in animals, pyruvate is converted to lactate. In plants and microorganisms such as yeast, it is converted to ethanol and carbon dioxide. Both of these pathways, summarised in Figure 5.18, involve the conversion of reduced NAD to NAD and allow glycolysis to continue.

**Figure 5.18**
Anaerobic respiration allows organisms to produce ATP even in the absence of oxygen

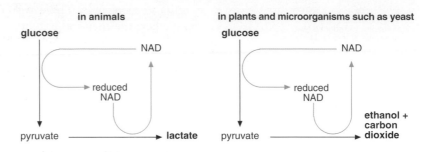

**Figure 5.19**
Insects such as this locust are able to fly long distances. When they start flying, their main respiratory substrate is glucose. After they have been in flight for 15 minutes or so, it is triglyceride

## Using different respiratory substrates

A **respiratory substrate** is the organic substance which forms the starting point for respiration. So far we have only looked at the respiratory pathway involving glucose as a respiratory substrate but other substances can also be respired (Figure 5.19)

One way in which it is possible to get some idea of an organism's respiratory substrate is to calculate its **respiratory quotient** (RQ). RQ can be calculated by dividing the amount of carbon dioxide produced by the amount of oxygen consumed in the same time. This ratio is different for different respiratory substrates so it can be used to indicate the substance which is being respired. The calculation in Extension box 2 shows how we can calculate the expected respiratory quotient for a triglyceride.

**Extension box 2**

### RQ calculations

The chemical equation below represents the respiration of a triglyceride

$$2C_{51}H_{98}O_6 + 145O_2 \rightarrow 102CO_2 + 98H_2O$$

The equation for respiratory quotient is:

$$RQ = \frac{\text{Amount of carbon dioxide produced}}{\text{Amount of oxygen consumed}}$$

Using the chemical equation we can enter the figures for the amounts of carbon dioxide produced and the amount of oxygen consumed.

$$RQ = \frac{102}{145}$$

$$= 0.70$$

We can carry out similar calculations to that shown in Extension box 2 to find the expected respiratory quotients for carbohydrates and proteins. These are shown in Table 5.1. We have to be careful how we interpret these figures. An RQ of 0.9, for example could mean that an organism is using protein as its respiratory substrate, but it could also be respiring a mixture of triglycerides and carbohydrates.

| RQ | Respiratory substrate |
|---|---|
| 0.7 | Triglycerides |
| 0.9 | Amino acids and proteins |
| 1.0 | Carbohydrates |

Table 5.1
The RQs for some important respiratory substrates

**Q 9** An organism is using protein as a respiratory substrate. In one minute it consumes 5.4 cm$^3$ of oxygen. How much carbon dioxide would you expect it to produce?

**Q 10** How would you expect the RQ of the locust shown in Figure 5.19 to change over the first fifteen minutes of a flight?

## Extension box 3

## Brown fat

Figure 5.20
When the animal is hibernating, the body temperature of a dormouse falls from its normal level of about 35 °C to just above 0 °C. Brown fat plays a very important role in allowing a dormouse to regain its normal body temperature rapidly when it comes out of hibernation

White fat plays an important part in insulating the body of a mammal, and helps to limit heat loss. Brown fat is very different. For one thing, brown fat cells are packed with mitochondria. This provides a clue as to its function. Another clue is provided by looking at those organisms which have large quantities of brown fat. It is found only in mammals, in particular, in very young mammals such as human babies, and in those which hibernate during the coldest part of the year (Figure 5.20).

Brown fat is the only tissue in the body of a mammal whose sole function is to produce heat. The internal membranes of its mitochondria have a special adaptation. In mitochondria in other tissues, hydrogen ions pass back from the space between the two mitochondrial membranes into the matrix through pores. These pores are associated with the enzyme ATP synthetase. The electron transport chain is therefore linked to ATP production and most of the potential energy in the hydrogen ions is transferred to ATP. In the mitochondria of brown fat, however, this mechanism can be over-ridden. The hydrogen ions can flow back through channels which are not associated with ATP synthetase so the energy they release will not be used to form ATP. It will all be released as heat.

In most mammals, brown fat is found in the chest, particularly around the larger arteries. The heat produced by brown fat can, therefore, be rapidly distributed round the body.

## Energy transfer and ecosystems

### Food chains and food webs

In any ecosystem different organisms gain their nutrients in different ways. Green plants are the **producers**. They are able to produce organic molecules from carbon dioxide and water. They rely on photosynthesis to transfer energy from sunlight to chemical potential energy in organic molecules. The other organisms which make up the community rely either directly or indirectly on organic molecules produced by the plants. **Primary consumers** feed on plants; **secondary consumers** feed on primary consumers; and **tertiary consumers** feed on secondary consumers. Organisms which are not eaten eventually die. Another group of organisms, the **decomposers**, break dead tissues down and use the organic molecules which make up these tissues as a source of potential chemical energy. We can summarise all this as a simple diagram (Figure 5.21).

We often talk about food chains suggesting perhaps that we frequently encounter situations where animal **B** feeds only on plant **A**. In turn, animal **C** eats animal **B**, and animal **D** eats animal **C**. This hardly ever happens in nature. Food chains are generally linked with each other to form complex food webs. We will now look at a specific ecosystem – the patch of nettles we mentioned at the beginning of the previous chapter. Look at Figure 5.22 but bear in mind we have kept the food web that this represents as simple as possible. We have left out a number of different organisms and completely omitted the decomposers. Note also that:

- some organisms, such as the dark green bush cricket, feed at different trophic levels. They feed on nettle leaves and they also feed on various species of insect.

- some organisms feed on different foods when they are larvae and when they are adult. The larva of the peacock butterfly, for example, feeds on nettle leaves. The adult feeds on the nectar produced by various flowers.

**Q 11** **Sundew is an insectivorous plant. It has sticky hairs on its green leaves which trap and digest small insects such as aphids. What trophic levels does sundew occupy?**

### Ecological pyramids

Diagrams such as Figure 5.22 only provide qualitative information. They only show what food different organisms eat. Biologists are interested in quantitative information. This is where ecological pyramids are particularly useful.

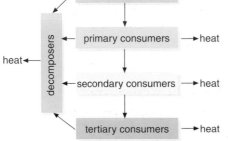

**Figure 5.21**
In this diagram, which shows the transfer of energy in an ecosystem, the boxes represent trophic levels. The arrows show the direction in which energy is transferred

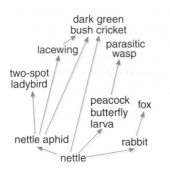

**Figure 5.22**
A simplified food web showing some of the organisms which feed on nettles

## Pyramids of numbers

The simplest way to compare the different trophic levels in an ecosystem, is to count the organisms present. We can represent this information as a pyramid of numbers. Three typical pyramids of number are shown in Figure 5.23. If we take the pyramid representing the food chain:

$$\text{nettle} \rightarrow \text{rabbit} \rightarrow \text{fox}$$

you will see that it is obviously pyramid shaped. If you count all the nettle plants, there are a lot more than the total number of rabbits. In turn there will be more rabbits than there are foxes There are two important exceptions to this pyramid shape when we are talking about numbers of organisms. The pyramid will be upside-down or inverted if a lot of small animals are feeding off a large plant. This is the case in the second example where the primary consumers are aphids. A second example where we get an inverted pyramid is where an organism is supporting a large number of small parasites such as humans and head lice or, in our example in Figure 5.23, parasitic wasps feeding on the caterpillars of peacock butterflies.

**Figure 5.23**
These examples of pyramids of numbers and pyramids of biomass all refer to food chains based on nettle plants

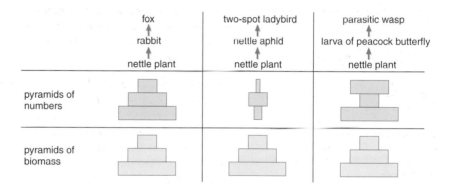

## Pyramids of biomass

Biomass is a measure of the total amount of living material present. If we shook all the aphids from a clump of nettle plants and weighed them we would have the biomass of the aphids on that clump of nettles. A pyramid of biomass allows us to compare the mass of organisms present at each trophic level of a particular food chain or food web. Pyramids of biomass overcome the problem of trying to compare numbers when the organisms are of very different size. Because of this, all the pyramids of biomass shown in the table have a similar pyramid shape. There is one important exception to this – organisms such as the phytoplankton shown in Figure 5.24 which multiply very rapidly. At any one time, their total biomass may be less than the biomass of the primary consumers which feed on them.

## The transfer of energy

Earlier in this chapter we considered the efficiency with which energy is transferred to plants in photosynthesis. In this section we will look at the efficiency with which energy is transferred to the consumers in an

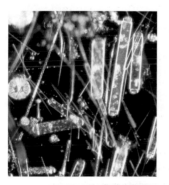

**Figure 5.24**
Phytoplankton are tiny single-celled algae. They are producers and float in the surface waters of lakes and oceans. Under favourable conditions they multiply very rapidly. Because of this, their total biomass may be less than that of the primary consumers. However, if we looked at their biomass over a period of say a month, then it would be greater than that of the primary consumers

ecosystem. Look at Figure 5.25. It shows the percentage of energy transferred between different trophic levels. We can look at this in another way. For every 10 000 kJ of light energy absorbed by the producers, 200 kJ will be incorporated into their tissues; 20 kJ will be transferred to the tissues of the primary consumers; and 2 kJ will end up as chemical potential energy in the tissues of secondary consumers. The numbers in this diagram are only generalisations. There are many factors which influence exactly how much energy is transferred at each stage. Some of these are considered in Extension box 4.

**Figure 5.25**
Only a small proportion of energy is transferred from one trophic level to another. The rest is used to make ATP in the process of respiration and will ultimately be lost as heat. Transferring only a small proportion limits the number of trophic levels in food webs. In practice it is rare for there to be more than five

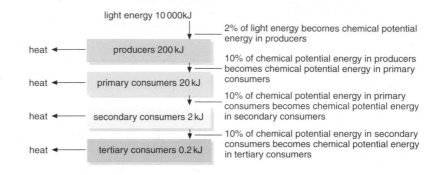

**Q 12** What percentage of the chemical potential energy in the tissues of a primary consumer would you expect to be transferred to the tissues of a tertiary consumer?

We can construct a third type of ecological pyramid which allows us to compare the amount of energy passing through the different levels in an ecosystem over a given period of time. This is called a pyramid of energy. Pyramids of energy are always pyramid-shaped. There are no exceptions.

## Extension box 4 — Energy transfer and efficiency

The figure we gave for the transfer of energy from producers to primary consumers and between primary consumers and secondary consumers was 10%. This figure, however, is only an average value. The actual proportion transferred varies from food chain to food chain. In this box we will look at some of the factors which help to determine the efficiency of energy transfer.

Mammals are homeothermic. This means that they are able to keep their body temperatures more or less constant at a value somewhere between about 35 °C and 40 °C, depending on the species. This high temperature is a result of heat produced during metabolism. Monitor lizards like that shown in Figure 5.26 are found in many parts of the tropics. They rely on their environment to maintain a high body temperature. More of the food they eat can therefore be converted into new cells and tissues and less chemical potential energy will be lost in maintaining body temperature.

**Figure 5.26**
Monitor lizard

**Figure 5.27**
Palm squirrel

**Figure 5.28**
Lion

You should remember from your AS course that the surface area to volume ratio of a small mammal such as the palm squirrel shown in Figure 5.27 is much bigger than that of a large mammal. Small mammals therefore lose a lot more heat relative to their size and cannot convert as much of the food they eat into new cells and tissues.

In general carnivores, like the lion in Figure 5.28, convert the food they eat into new tissue more efficiently than do herbivores. Herbivores feed on plant material and plants contain a lot of substances such as cellulose and lignin which are difficult to digest. A much higher proportion of the food that a herbivore eats passes through the gut and is lost as faeces.

**Q 13** Which do you think would be more efficient at transferring chemical potential energy in food into chemical potential energy in its tissues – a young animal or a mature animal? Explain your answer.

So which animal converts food into new tissue most efficiently? The answer is probably trout living in fish farms. They can convert as much as 20% of the food they eat into new tissue. The information in this box will help to explain why. Trout are poikilotherms so do not need to use much of their food to maintain body temperature; they are carnivores and there is little waste from the food pellets on which they are fed; and they are harvested while they are still young and growing rapidly.

## Summary

- Photosynthesis uses energy from sunlight to synthesise organic molecules from carbon dioxide and water. The biochemical reactions involved can be divided into the light dependent and light-independent reactions.

- Respiration is a biochemical process in which organic substances, called respiratory substrates, are broken down in a series of stages and the chemical potential energy they contain is transferred to ATP.

- ATP is an immediate source of energy for various forms of biological work such as movement, the synthesis of large organic molecules and active transport.

- Energy is transferred from one trophic level to another through food chains and food webs in a community.

**Figure 5.29**
A selection of different lichens. Each lichen consists of a fungus and an alga. The lichen fungi are unable to live without their algal partners. Many of the algae, however, can be found free-living

# Assignment

## Sharing a life

A lichen, such as one of those shown in Figure 5.29, is made up of two entirely different organisms. It is a partnership of an alga and a fungus. When a section through a lichen is examined with a microscope, green algal cells are clearly seen among thread-like fungal hyphae.

Many lichens are able to withstand extreme conditions. Some species colonise and grow on bare rock and must be able to survive for long periods without water. Others live in the Arctic or in deserts. Despite this, they are extremely sensitive to pollution. In this assignment we will look at the way in which lichens are affected by sulphur dioxide in the atmosphere. In order to answer the questions, you will need to draw on your knowledge of photosynthesis as well as your understanding of ecology.

Read the following passage about the relationship between the algae and the fungus which make up a lichen.

> Nutrition in lichens has been investigated with carbon dioxide containing the radioactive isotope of carbon, $^{14}$C. The lichen is incubated with radioactive carbon dioxide. The total amount of carbon dioxide fixed by the algal cells and the total amount
> 5   transferred to the fungus are then calculated. The results show that the total amount transferred is between 60 and 80%. These results may be misleading, however. First, it is very difficult to separate the algal and fungal partners. Second, some of the photosynthate transferred to the fungus is used as a respiratory substrate.
>
> 10   The substance transferred from the algae varies from one lichen to another. Once in the fungus, however, it is used for respiration or to form other substances. A lot of the photosynthate taken up by fungi is converted to substances called polyols. Polyols are soluble and have a marked effect on the water potential of the fungal hyphae.
> 15   This enables them to absorb water rapidly.

1   Explain the meaning of the following terms used in the passage:

    (a)  fixed (line 4)

    (b)  photosynthate (line 8)

                        *(2 marks)*

2   Explain how each of the following contributes to the results of these investigations being unreliable:

    (a)  the difficulty of separating the algal and fungal partners (lines 7–8);

                        *(2 marks)*

(b) lichens using photosynthate as a respiratory substrate
(lines 9–10).

*(2 marks)*

3 Lichens have no means of controlling water loss. They dry out
readily and remain inert until they are wetted again. Explain, in
terms of water potential, how the presence of polyols enable lichens
to absorb water rapidly when they are wetted.

*(2 marks)*

Lichens are very sensitive to the presence of sulphur dioxide. The
main effect of this pollutant is on photosynthesis in the algal partner.
Increasing the concentration of sulphur dioxide produces a series of
increasingly severe effects.

These are:

● a temporary reduction in the rate of photosynthesis which recovers
as soon as sulphur dioxide concentration is reduced.

● a permanent reduction in the rate of photosynthesis. This is
probably the result of damage to chloroplast membranes inhibiting
ATP production.

● a permanent reduction in the rate of photosynthesis related to the
breakdown of chlorophyll.

4 (a) Explain why damage to chloroplast membranes may inhibit
ATP production.

*(2 marks)*

(b) Explain how a lack of ATP will affect the light-independent
reaction of photosynthesis.

*(3 marks)*

5 Describe briefly how you could use chromatography to investigate
whether chlorophyll had been broken down in lichens which had
been exposed to very high concentrations of sulphur dioxide.

*(4 marks)*

It is thought that lichens are very susceptible to sulphur dioxide
pollution because they have a low proportion of chlorophyll-
containing tissue. Experimental work has shown that different species
of lichen are affected to different extents by sulphur dioxide. Table 5.2
shows the effect of different concentrations of sulphur dioxide on the
rate of uptake of carbon dioxide in three species of lichen:

| Sulphur dioxide concentration/ arbitrary units | Rate of uptake of carbon dioxide by lichen/arbitrary units | | |
|---|---|---|---|
| | *Usnea subfloridana* | *Hypogymnia physodes* | *Lecanora conizaeiodes* |
| 0 | 300 | 330 | 280 |
| 0.1 | 320 | 440 | |
| 0.2 | 210 | 340 | 320 |
| 0.3 | 30 | 390 | |
| 0.4 | 0 | 0 | 280 |
| 0.6 | | | 40 |
| 0.8 | | | 10 |

Table 5.2
Effect of different concentrations of sulphur dioxide on the rate of uptake of carbon dioxide in three species of lichen

6   Plot these data as a suitable graph on a single pair of axes.

*(4 marks)*

7   Suggest how sulphur dioxide may affect the distribution of these three species of lichen.

*(2 marks)*

# Examination questions

1   The diagram shows the flow of energy through a marine ecosystem. The units are kJ m$^{-2}$ year$^{-1}$.

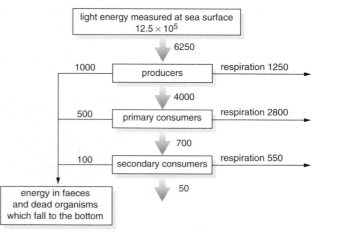

(a) (i)  Calculate the percentage of the light energy at the sea surface which is converted into chemical potential energy in the producers. Show your working.

*(2 marks)*

(ii) The percentage of the light energy at the sea surface which is converted into chemical potential energy in the producers is very small. Give **two** reasons for this.

*(2 marks)*

(b) Use the information in the diagram to explain why marine ecosystems such as this rarely have more than five trophic levels.

*(2 marks)*

(c) What happens to the energy in faeces and dead organisms which fall to the bottom of the sea?

*(2 marks)*

(d) Light energy is important in the light-dependent reaction of photosynthesis. The energy changes which take place in the light-dependent reaction are shown in the diagram.

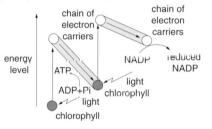

(i) Describe what happens to the chlorophyll when it is struck by light.

*(2 marks)*

(ii) The weedkiller DCMU blocks the flow of electrons along the chains of electron carriers. Describe and explain the effect this will have on the production of triose phosphate in the light-independent reaction.

*(3 marks)*

(e) Living organisms release energy from organic molecules such as glucose during respiration. Much of this energy is used to produce ATP. Explain why ATP is better than glucose as an immediate energy source for cell metabolism.

*(2 marks)*

(f) The production of ATP is said to be coupled to the transport of electrons along the carrier chain. Normally, electrons are only passed along the electron chain if ADP is being converted to ATP at the same time. When the amount of ADP in a cell is low, electrons do not flow from the reduced coenzyme to oxygen.

(i) Suggest how the rate of respiration is linked to the needs of the cell.

*(3 marks)*

(ii) DNP is a substance which allows electron transport to take place without the production of ATP. When DNP is given to rats, their body temperatures rise. Explain why.

*(2 marks)*

2   A suspension of pure chloroplasts was prepared from plant cells by cell fractionation and centrifugation. Some of these chloroplasts were broken up and used to produce a suspension of chloroplast membranes. The table shows the results of a number of investigations with the intact chloroplasts and with the chloroplast membranes.

| | Intact chloroplasts | | Chloroplast membranes | |
| --- | --- | --- | --- | --- |
| | light | dark | light | dark |
| Volume of oxygen produced/mm$^3$ hour$^{-1}$ | 480 | 0 | 400 | 0 |
| Amount of carbon dioxide taken up/arbitrary units | 52 000 | 238 | 40 | 37 |
| Amount of inorganic phosphate taken up/ arbitrary units | 4 250 | 950 | 1 100 | 300 |

(a) (i)  Describe how you could obtain a preparation of chloroplasts which was not contaminated with cell nuclei.

*(2 marks)*

(b) (i)  Describe what happens to the inorganic phosphate taken up by the chloroplasts.

*(1 mark)*

(ii) Explain the evidence from this table that the light-independent reaction takes place in the stroma of the chloroplasts.

*(2 marks)*

3  (a)  The inner membrane of a mitochondrion is folded to form cristae. Suggest how this folding is an adaptation to the function of a mitochondrion.

*(2 marks)*

(b)  The equation represents oxidation of a lipid.

$$C_{57}H_{104}O_6 + 80O_2 \rightarrow 57CO_2 + 52H_2O + \text{Energy}$$

Use the equation to calculate the Respiratory Quotient (RQ) of this lipid. Show your working.

*(1 mark)*

(c)  Suggest an explanation for each of the following:
(i)  the RQ of germinating maize grains is 1;

*(1 mark)*

(ii) the RQ of a normal healthy person varies over a 24-hour period.

*(1 mark)*

# Decomposition and Recycling

All too often summer holidays in Cornwall are spoilt by rain. And when it rains, tourists need somewhere to go. One popular place is the Elizabethan manor house at Trerice, now owned by the National Trust. Several hundred people may visit the house and its teashop in a single day and most of them will use the toilets and washrooms. Unfortunately neither the house nor the teashop is connected to mains drainage.

Waste runs first of all to a septic tank. Here, the solid material separates out to form sediment at the bottom of the tank. The liquid is then pumped into a reed bed (Figure 6.1) where it is purified further. It flows through the sandy soil in which the reeds are growing. There are huge numbers of microorganisms around the roots of the reeds and these act on the suspended organic substances in the incoming water. They break these substances down and release soluble nitrates and phosphates as a result. Some nitrate and phosphate is contained in the water leaving the reedbed but a lot of it is taken up by the plants. In their cells it is converted into organic substances such as proteins and lipids. The water that flows out of the reed bed is a lot cleaner than that flowing in. There is much less suspended organic material and a lower concentration of ions such as nitrates and phosphates.

Figure 6.1
A reed bed

Reeds are tall grasses which grow at the edge of lakes and rivers where they may form large swampy areas. Not only are the resulting 'wetlands' important habitats for wildlife, but they can also be used to help deal with certain types of pollution.

In chapter 5 we saw how energy was transferred from one trophic level in a food web to the next. Transfer of energy is inefficient and it is all eventually lost as heat. Energy is described as a renewable resource. Ions like the nitrates and phosphates in the waste from Trerice are, however, non-renewable. There is only a certain amount of them and, like all nutrients, they are recycled. In this chapter, we shall look at the passage of carbon and nitrogen through the various trophic levels in an ecosystem and at the role of microorganisms in converting organic molecules into inorganic substances which are made available to plants.

## Nutrient cycles

Look at Table 6.1 overleaf. You will see that living organisms such as animals and plants require different chemical elements. Plants take these elements up either as carbon dioxide in the case of carbon, or as ions from the soil. Inside the plant, they are involved in various chemical reactions and are eventually converted into the organic substances which form plant cells and tissues. Consumers obtain their supply of these elements from plants or from other animals which feed on plants.

An insect, for example, digests the complex organic molecules in its food and absorbs the products through its gut wall. In this way, elements are passed from organism to organism along the food chains which make up a food web (Figure 6.2).

Those organisms or parts of organisms which are not eaten as food will eventually die and decompose. Microorganisms such as fungi and bacteria break down the complex organic molecules which form the dead material. They absorb some of the products but the rest are released as inorganic substances which can be taken up by plants.

So, here we have the basic nutrient cycle. Elements are taken up by plants and passed from organism to organism in the various trophic levels. Death and decomposition result in microorganisms making these elements available to plants again.

| Chemical element | Plants | Animals |
| --- | --- | --- |
| Carbon | The essential component of all organic substances. Proteins, carbohydrates, lipids and nucleic acids are all built up from carbon-containing molecules. Many other substances such as ATP and chlorophyll also contain carbon. | |
| Nitrogen | An essential component of all amino acids, and of the nucleotides which make up DNA and RNA. | |
| Iron | Contained in the cytochromes which form part of the electron transport chain. Needed for enzymes such as catalase to work. Essential for the synthesis of chlorophyll. | Contained in the cytochromes which form part of the electron transport chain. Needed for enzymes such as catalase to work. Part of haemoglobin. |
| Iodine | | Contained in thyroxine. This is a hormone produced by the thyroid gland. |
| Molybdenum | Needed for the enzyme nitrate reductase to work. This enzyme reduces nitrates during the synthesis of amino acids. | |

**Table 6.1**
Some examples of chemical elements required by plants and animals

**Figure 6.2**
The basic nutrient cycle. All elements, whether they are carbon or nitrogen, iron or molybdenum, are cycled in this basic way. It is only the details of the process that differ in different cycles

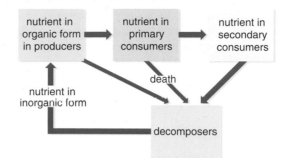

**Q** 1 **Using the basic cycle shown in Figure 6.2 as a guide, draw a simple diagram representing an iron cycle.**

## The carbon cycle

Plants take up carbon dioxide from the atmosphere. It diffuses into the photosynthesising cells in the leaves where it is converted to sugars and other organic compounds. Primary consumers feed on plants, and carbon-containing compounds in their food are digested. Smaller, soluble molecules are produced and these are absorbed. They are built up into the carbohydrates and other large molecules which make up the tissues of the primary consumers. The same thing happens when a primary consumer is eaten by a secondary consumer. In this way, just as in other nutrient cycles, the element is passed from one trophic level to the next through the food web.

The next step involves the decomposers. Many soil bacteria and fungi are saprobionts. They secrete enzymes which break down the complex organic substances which make up dead plants and animals. The smaller molecules which result are absorbed by the microorganisms and some of them are used in respiration. Respiration releases carbon dioxide which can be used by plants for photosynthesis. The cycle (Figure 6.3) is complete.

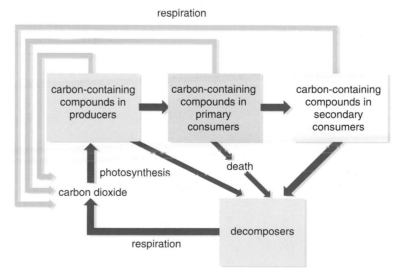

**Figure 6.3**
The carbon cycle. Note how similar this cycle is to the basic nutrient cycle shown in Figure 6.2

However, there is one complicating factor which we do not see in other cycles. We do not have to wait for an animal or plant to die before any of the carbon stored in its tissues is released. The producers and consumers all respire and release carbon dioxide into the atmosphere.

**Q** 2 In which of the following ways does a maize plant take up carbon?
   A as carbon dioxide in the air
   B as hydrogencarbonate ions in the soil
   C as carbonate ions in the soil
   D as organic molecules from dead plants and animals

**Figure 6.4**
Graph showing how the concentration of carbon dioxide in a sugar beet crop varies. The blue circles (●) show the mean values at night time (20–04h) and the red circles (●) show mean values during the daylight (08–16h). The lines show the overall trends

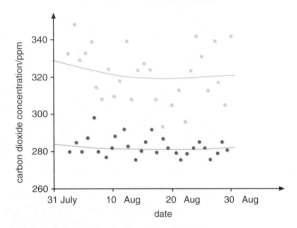

## Respiration, photosynthesis and carbon dioxide

The concentration of carbon dioxide in the atmosphere is not constant. It changes according to the balance between photosynthesis and respiration. Look at the graph in Figure 6.4. It shows how the concentration of carbon dioxide varied in a sugar beet crop in which the plants had reached their full height. All the measurements were made at crop height, 50 cm above the ground.

This graph shows us that the concentration of carbon dioxide at crop height is greater at night time than during daylight. There is a simple explanation for this difference. At night, plants are unable to photosynthesise but they are still respiring. During the daytime, both photosynthesis and respiration take place. Since the rate of photosynthesis is usually greater than the rate of respiration, more carbon dioxide will be taken up than is released and its concentration falls.

If we look at the graph more carefully, there is another feature that we might notice. There is a lot of variation in the mean carbon dioxide concentration at night. The points are scattered quite widely. Much of this variation is due to the wind. On still nights, carbon dioxide accumulates round the plants and is not mixed with the surrounding air, so values are higher. On windy nights, it is mixed with air from higher up resulting in lower values.

## Human activities and the carbon cycle

So far we have only looked at the biological aspects of the carbon cycle. We have ignored the fact that large amounts of carbon are locked up in coal, oil and limestone formed millions of years ago. The ways in which human activities influence the carbon cycle are nearly all involved with releasing this 'stored' carbon as carbon dioxide. These include the burning of fossil fuels such as coal and oil (Figure 6.5) and the weathering of limestone (Figure 6.6).

We will look at one other human activity in a little more detail, the clearing of forests. In a rain forest such as that shown in Figure 6.7, there is a balance between the rate of photosynthesis and the overall rate of respiration. The carbon dioxide removed from the atmosphere in photosynthesis is roughly equal to the amount put back in by the respiration of all the plants, animals and microorganisms that make up the ecosystem.

Forest may have been growing in a particular area for hundreds of years and, in that time, a lot of carbon has been locked up in the tissues of the trees and other plants as compounds like cellulose and lignin. If we want to clear the land for agriculture, we need to get rid of the forest and the easiest way to do this is by burning. Burning releases the carbon locked in the tissues into the atmosphere as carbon dioxide.

The result of all of these human activities is that there has been a steady rise in the atmospheric concentration of carbon dioxide. Figure 6.8 provides one piece of evidence for this. It shows a graph of the concentration of carbon dioxide in the atmosphere at the Mauna Loa observatory in Hawaii. Records have been kept here for over 50 years but, to see the detail, we will look at a ten-year period.

**Figure 6.5**
These vapour trails were left by aircraft. It is not only our increasing reliance on motor vehicles which increases the rate at which we burn fossil fuels but also increased consumption by aircraft and power stations

**Figure 6.6**
The limestone from which these houses were built is a rock formed from the skeletons of tiny marine organisms. It consists mainly of calcium carbonate. Using limestone for building increases the surface area exposed to the weathering processes which produce carbon dioxide

**Figure 6.7**
An area of mature rainforest like this has a neutral effect on the concentration of carbon dioxide in the atmosphere. The amount of carbon dioxide removed in photosynthesis is more or less the same as the amount replaced by respiration

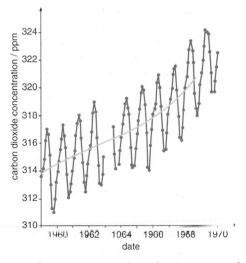

**Figure 6.8**
The individual points on this graph show the mean monthly carbon dioxide concentrations at the Mauna Loa observatory in Hawaii. The curve shows the general trend in the mean annual carbon dioxide concentration

In order to interpret these figures we need to be aware that the gases in the atmosphere are continually being mixed by wind and air currents. This means that the graph reflects the carbon dioxide concentration of the atmosphere over the whole of the northern hemisphere, not just over Hawaii. You will see that there is an annual fluctuation in carbon dioxide concentration. The peaks are in the winter months. There are two main reasons for this. First, oil consumption is much higher at this time of the year and, second, the overall rate of photosynthesis is lower during the winter.

The second aspect of this graph is that there is an overall trend towards an increase in the mean annual concentration of carbon dioxide in the atmosphere. This is a cause of considerable concern as increase in carbon dioxide has been linked to global warming and its consequences.

**Q** 3 Would you expect the trend in atmospheric carbon dioxide concentration between 1990 and 2000 to have been the same, greater or less than in the period shown in the graph? Give a reason for your answer.

## Extension box 1

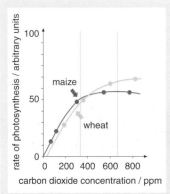

**Figure 6.9**
The effect of carbon dioxide concentration on the rate of photosynthesis in wheat and maize. Compare what happens to these two crops if we double the atmospheric carbon dioxide concentration from 330 ppm to 660 ppm

## Increasing carbon dioxide – a good thing?

Most people link an increase in the carbon dioxide concentration in the atmosphere with global warming, rising sea levels and climatic change. But could an increase in carbon dioxide concentration have a beneficial effect on the growth of crop plants? This is not a simple question to answer. An increase in carbon dioxide concentration is likely to be accompanied by a series of other changes and these separate factors will interact with each other in a way that is difficult to predict.

Plants depend on carbon dioxide for photosynthesis. If you followed the AS Biology course, you will probably remember that in bright conditions it is the concentration of carbon dioxide that usually limits the rate of photosynthesis. Increasing carbon dioxide concentration should therefore increase photosynthetic rate. Look at the graph in Figure 6.9. It shows the effect of an increase in carbon dioxide concentration on photosynthesis in two important cereal crops, wheat and maize.

Wheat is a $C_3$ plant. This means that it uses the biochemical pathway we described in Chapter 5 for photosynthesis. Many of the important food crops grown in temperate regions, crops such as rice and soya beans, are also $C_3$ plants. They respond to an increase in carbon dioxide concentration in a similar way to wheat. An increase in the concentration of carbon dioxide in the atmosphere ought then to result in a significant increase in food production.

Some plants, however, photosynthesise using a slightly different biochemical pathway, the $C_4$ pathway. These plants include important tropical plants such as maize and sorghum. Increasing carbon dioxide concentration only brings about a small increase in their rate of photosynthesis. In practice, this slight increase may not be enough to offset decreases brought about by other factors associated with increased carbon dioxide concentration and global warming.

So far the story is quite simple. Increased carbon dioxide concentration may result in a significant increase in yield in $C_3$ plants; it is likely to produce a much smaller one in $C_4$ plants. There are, however, complicating factors. Only a certain amount of the carbohydrates and other substances produced by photosynthesis go into the part of the plant that is eaten. Look at the figures in Table 6.2. All these data refer to $C_3$ plants and predict the effect of doubling the present concentrations of carbon dioxide.

| Crop | Percentage increase in biomass | Percentage increase in marketable yield |
|---|---|---|
| Cotton | 124 | 104 |
| Tomato | 40 | 21 |
| Rice | 20 | 36 |
| Cabbage | 37 | 19 |
| Weeds | 34 | not applicable |

**Table 6.2**
Effects of doubling the present concentrations of carbon dioxide on selected $C_3$ plants

In some of these crops, such as tomatoes, you can see that the increase in the yield of the part of the plant that we eat is nowhere near as high as the increase in total plant growth. It is worth noting as well that an increase in carbon dioxide concentration will increase weed growth.

What about crop quality? More does not necessarily mean better and there is evidence to suggest that plants grown under conditions where the carbon dioxide concentration is higher have a relatively lower proportion of nitrogen, and hence protein, in their tissues.

So, in terms of crop yield, is increased carbon dioxide concentration a good thing? We simply do not have the evidence to answer this complex question.

**Q** 4 **Give two reasons why an increase in carbon dioxide concentration may lead to a farmer gaining a smaller profit from a crop.**

## The nitrogen cycle

The general features of the nitrogen cycle differ very little from the basic nutrient cycle shown in Figure 6.2. Plants take up nitrate ions from the soil. They are absorbed into the roots by active transport and used to produce proteins and other nitrogen-containing compounds in plant cells. Primary consumers feed on plants and the proteins in their food are digested to form amino acids. The amino acids are absorbed from the gut and built up into the proteins which make up the tissues of the primary consumers. The same thing happens when a primary consumer is eaten by a secondary consumer. In this way nitrogen is passed from one trophic level to the next through the food web.

When organisms die, the nitrogen-containing organic substances which they contain are digested by saprobiotic bacteria and ammonia is released. Nitrogen from consumers is also made available to saprobiotic bacteria through excretory products such as urea. Another group of bacteria, the nitrifying bacteria, then convert ammonia to nitrites and nitrates. This complete nitrogen cycle is summarised in Figure 6.10.

**Figure 6.10**
Note how similar this basic nitrogen cycle is to the basic nutrient cycle shown in Figure 6.2 and to the carbon cycle shown in Figure 6.3

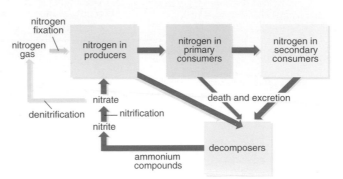

Under anaerobic conditions, such as those that are found when soil becomes waterlogged (Figure 6.11), **denitrifying bacteria** are found in large numbers. These bacteria are able to use nitrate instead of oxygen as an electron acceptor in the respiratory pathway. The reaction involves reduction of nitrate to nitrogen gas. This nitrogen escapes into the atmosphere so it is no longer available to plants.

**Q** 5 **Give two differences between nitrification and denitrification.**

Nitrogen gas can be made available to plants again by **nitrogen fixation**. This term refers to a number of processes which involve the conversion of nitrogen gas to nitrogen-containing substances. These processes all require a considerable amount of energy. A nitrogen molecule consists of two atoms, and the molecules have to be split into these atoms before nitrogen fixation can take place. Energy is required for this. Methods of fixing nitrogen include:

### Lightning

During a thunderstorm, the electrical energy in lightning allows nitrogen and oxygen to combine to form various oxides of nitrogen. Rain washes these compounds into the soil where they can be taken up by plants as nitrates. In the UK, the amount of nitrogen fixed by lightning is small. In some parts of the tropics, however, violent thunderstorms are common and this is an important way of fixing nitrogen.

### Industrial processes

The Haber process combines nitrogen and hydrogen to produce ammonia. The reactions involved take place at high temperatures and pressures and require the presence of a catalyst. A lot of the ammonia produced by the Haber process is used to make fertilisers which are added to the soil.

**Figure 6.11**
When fields are flooded, the spaces between soil particles are filled with water. No oxygen is available for soil microorganisms. Under these waterlogged conditions, denitrifying bacteria are common

## Nitrogen-fixing microorganisms

Many microorganisms are able to fix nitrogen. Some of them live free in the soil. Others are associated with the roots of leguminous plants such as peas, beans and lupins (Figure 6.12).

The biochemistry of nitrogen fixation is very similar in all these organisms and is summarised in the equation:

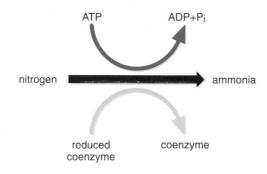

Nitrogen is reduced to ammonia. The reaction is catalysed by the enzyme nitrogenase. Nitrogenase, however, does not function in the presence of oxygen and many nitrogen-fixing microorganisms have adaptations which ensure that anaerobic conditions exist in the parts of the cells involved in nitrogen fixation.

**Figure 6.12**
The swellings on these roots are nodules. Inside each nodule are large numbers of nitrogen-fixing bacteria. The bacteria and the plant have a mutualistic relationship. The bacteria get their carbon-containing compounds from the plant while the plant gains ammonium ions from the bacteria

**Q** 6 What is the function of ATP in nitrogen fixation?

## Human activities and the nitrogen cycle

When we grow a crop such as wheat, we do not allow the plants to complete their lives and then die and decompose. We harvest them. Instead of being recycled, nitrogen-containing substances in the grain and the straw are taken away for our use. To make up this loss we need to add fertiliser. Nitrogen-containing fertilisers contain nitrate and ammonium ions. These ions are very soluble so they can be readily leached from the soil into lakes and rivers. You may have studied the environmental effects of this as part of the AS course but you should look at this material again. It links closely with the nitrogen cycle described in this chapter. Extension box 2 is reprinted from the AS book *A New Introduction to Biology* and should remind you of some of the important details.

**Extension box 2**    Where do all the ions come from?

**Figure 6.13**
The vigorous growth of algae and water plants in this East Anglian stream is the result of pollution with nitrate and phosphate ions

The stream shown in Figure 6.13, which runs through an area of farmland in East Anglia, is very heavily polluted. The high concentrations of nitrate and phosphate ions in the water have led to increased growth of algae and water plants. As the stream is in a rural area, little of this pollution comes from sewage. It is mainly the result of agriculture. There are two main sources of these polluting nitrates and phosphates:

● fertilisers add nutrients to the soil and not all of these are taken up by crop plants. The rain leaches soluble ions from the soil, with the result that they drain into streams and ponds.

● cattle and other livestock produce large amounts of urine and faeces. This material is not always disposed of efficiently and can lead indirectly to an increase in the concentration of nitrate and phosphate ions in streams such as that shown in Figure 6.13.

Figure 6.14 shows the monthly concentration of nitrate ions in this stream. It also shows the total volume of water present. This is indicated by the curve showing the flow.

**Figure 6.14**
Graph showing how nitrate ion concentration and stream flow vary at different times of the year

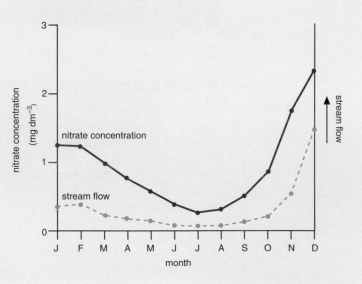

This is a relatively small stream so the flow provides a good indication of the amount of rain falling on the land immediately around it. During the winter months stream flow and rainfall are high. More rainwater drains into the stream during these months so more nitrates are leached from the surrounding farmland. Other factors are also involved. Crops sown in the autumn are only just beginning to grow so there is a lot of bare soil. Heavy rain results in water running off the soil, taking soil nutrients with it. It is cold and plants only grow slowly. Consequently, at this time of the year they take up a relatively small amount of mineral ions from the soil.

In the summer months things are different. Crop plants are photosynthesising and growing rapidly in the warmer conditions. Consequently, they remove nitrates from the soil, leaving less to be leached into the stream. Those ions in the stream are also being absorbed by the water plants as they grow.

Now look at Figure 6.15. This graph shows that there is a positive correlation between the concentration of nitrate in the stream and the amount of nitrate added in fertiliser to the surrounding land. The more nitrate added as fertiliser in a particular year, the more nitrate there is in the stream. We have to be careful about the conclusions that we draw from sets of data, however. We must be aware that just because these two things are correlated, it does not mean that the fertiliser is the only cause of pollution in the stream.

**Figure 6.15**
Graph showing the relationship between the mean concentration of nitrate in the stream and that added as fertiliser applied to the surrounding land

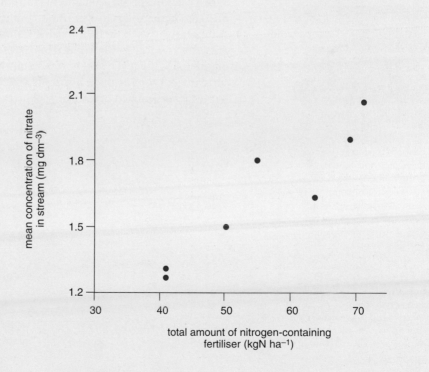

**Figure 6.16**
The cycling of nitrogen in a rain forest

The figures in boxes represent the amount of nitrogen contained in the soil and in the trees. The figures by the arrows show the flow of nitrogen over the course of a year. The units are kilograms per hectare

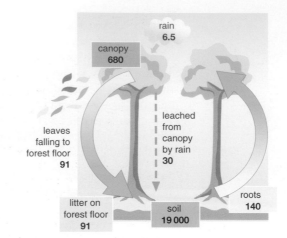

**Figure 6.17**
Fungi (a) are very common on the floor of the forest. Together with bacteria and animals such as giant millipedes (b), they break down the organic substances in fallen leaves and release nitrates which can be taken up by the forest plants

**Figure 6.18**
This graph shows the yield of grain from successive rice crops grown on an area where the forest has been cleared. Unless fertilisers are added, it soon becomes uneconomical to grow crops

We will look now at another way in which human activities influence the cycling of elements such as nitrogen – the clearing of forests. Look at Figure 6.16. It shows the distribution of nitrogen in the soil and the trees which make up a rainforest. It used to be thought that the soil on which the forest grew was very poor in nutrients; most of the nitrogen was assumed to be locked up in the trees. We now know this is incorrect. There is a lot of nitrogen in the soil, but most of this is in forms which the trees are unable to use. What is true, however, is that nitrogen in dead animal and plant material is recycled very rapidly (Figure 6.17).

When the forest is cleared and burnt to make way for crops, a lot of the nitrogen in the soil is lost in smoke but the ash that remains is still comparatively rich in the nutrients required for crops. The first crops grown on cleared land produce high yields. If you look at the graph in Figure 6.18, however, you will see that the yield rapidly falls with subsequent crops as these nutrients are either removed as the crops are harvested or leached out of the soil by rain.

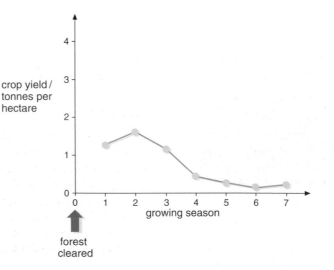

# Summary

- Nutrients are examples of non-renewable resources and are constantly recycled. They are taken up by plants and passed from organism to organism in the various trophic levels. Death and decomposition result in microorganisms making them available to plants again.

- Carbon cycles follow the general pattern shown by other nutrient cycles. Carbon, however, is taken up by plants as carbon dioxide rather than as an ion from the soil, and carbon dioxide is returned to the air by respiration.

- Nitrogen cycles also follow the general pattern shown by other nutrient cycles. Nitrogen passes in organic compounds from organism to organism in the various tropic levels. Saprotrophic bacteria break down dead organic material and release ammonium ions. Nitrifying bacteria release nitrates from these ammonium ions. The nitrates are taken up by plants

- Some nitrates are converted to nitrogen gas in the process of denitrification. Nitrogen gas can be made available to plants again by nitrogen fixation.

- Human activities affect the cycling of both carbon and nitrogen. The burning of fossil fuels increases the concentration of carbon dioxide in the atmosphere and is one of the main factors contributing to global warming. Nitrate ions are soluble and may leach into freshwater where their accumulation eventually leads to a fall in oxygen concentration and the death of many organisms.

- The clearing and burning of rain forest releases carbon dioxide from the carbon stored in forest trees. Nutrients released from the soil are rapidly leached out by rain or removed when crops are harvested.

**Figure 6.19**
Hedges can be thought of as long, narrow strips of woodland. They form a habitat for animals such as this wood mouse

# Assignment

## Hedges

Most of the United Kingdom was once covered in forest. Much of this natural woodland has now disappeared, cleared by an expanding human population for agriculture and industry. Many woodland plants and animals now find the only suitable habitat, the hedges which surround our fields (Figure 6.19). In this assignment we will look at some of the arguments for and against removing hedges. In order to answer the questions you will need to make use of some of the information from Chapter 4 as well that from this chapter.

The maps in Figure 6.20 are drawn from aerial photographs. They show the hedges around fields in one area in the Midlands in 1946 and in the same area in 1965.

**Figure 6.20**
Maps showing hedges in an area of the Midlands in (a) 1946 and (b) 1965

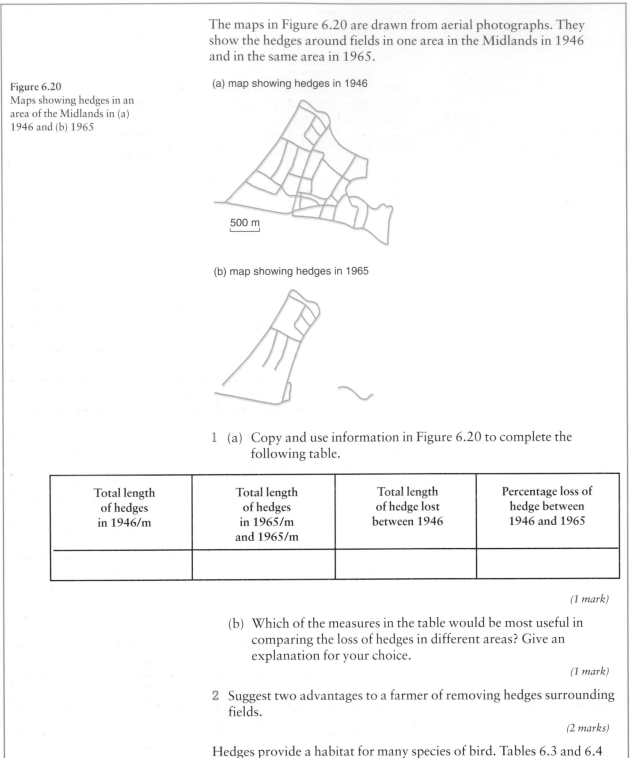

(a) map showing hedges in 1946

500 m

(b) map showing hedges in 1965

1  (a)  Copy and use information in Figure 6.20 to complete the following table.

| Total length of hedges in 1946/m | Total length of hedges in 1965/m and 1965/m | Total length of hedge lost between 1946 | Percentage loss of hedge between 1946 and 1965 |
|---|---|---|---|
|  |  |  |  |

*(1 mark)*

(b)  Which of the measures in the table would be most useful in comparing the loss of hedges in different areas? Give an explanation for your choice.

*(1 mark)*

2  Suggest two advantages to a farmer of removing hedges surrounding fields.

*(2 marks)*

Hedges provide a habitat for many species of bird. Tables 6.3 and 6.4 show the results of a study into the effects of different ways of managing hawthorn hedges on the species of birds found in them.

| Type of field boundary | Total length/m | Notes on management |
|---|---|---|
| A  No hedge present | 4130 | Not applicable |
| B  Remnant | 2320 | Hedge neglected. A few isolated bushes and small trees with large gaps in between are all that remains of the original hedge. |
| C  Mechanically cut | 2805 | Mechanically cut with tractor and hedge trimmer once a year |
| D  Overgrown | 2645 | Hedge not managed in any way. |

Table 6.3
Hedges in the study area and their management

| Species | Pairs of breeding birds in type of field boundary | | | |
|---|---|---|---|---|
| | A | B | C | D |
| Red-legged partridge | 6 | 2 | 3 | 1 |
| Skylark | 4 | 0 | 0 | 0 |
| Corn bunting | 5 | 4 | 0 | 1 |
| Whitethroat | 2 | 2 | 5 | 6 |
| Yellow hammer | 0 | 0 | 6 | 6 |
| House sparrow | 0 | 1 | 0 | 2 |
| Song thrush | 0 | 1 | 0 | 2 |
| Blackbird | 0 | 0 | 4 | 11 |
| Linnet | 0 | 0 | 3 | 6 |
| Bullfinch | 0 | 0 | 0 | 2 |
| Dunnock | 0 | 0 | 4 | 5 |
| Chaffinch | 0 | 0 | 1 | 3 |
| Goldfinch | 0 | 0 | 0 | 1 |
| Blackcap | 0 | 0 | 0 | 1 |
| Wren | 0 | 0 | 0 | 3 |

Table 6.4
Breeding birds present in different types of field boundary

3 How do the different methods of management affect the number of species and the diversity of breeding birds?

*(3 marks)*

Hedges affect the abiotic environment of crops growing in the fields they surround. Figure 6.21 and Table 6.3 show some of the results of an investigation of these effects.

**Figure 6.21**
The effect of a hedge on crop yield. In this area water shortage normally limits crop yield

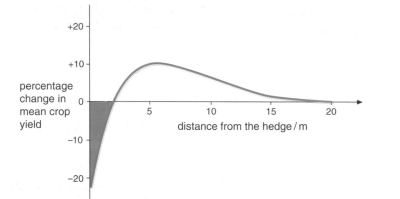

| Factor | Effect of hedge as percentage of value in the open at different distances from the hedge | | | | |
|---|---|---|---|---|---|
| | 2 m | 4 m | 8 m | 12 m | 16 m |
| Soil moisture | 118 | 118 | 110 | 104 | 100 |
| Daytime air temperature | 114 | 112 | 109 | 102 | 100 |
| Evaporation | 74 | 65 | 78 | 90 | 97 |
| Wind speed | 61 | 39 | 52 | 65 | 80 |

**Table 6.5**
The estimated effect of a hedge on various abiotic factors

4 The mean yield for the crop in this investigation was 5.3 tonnes hectare$^{-1}$. Use Figure 6.18 to calculate the mean yield of this crop:

(a) 2 metres from the hedge

(b) 7 metres from the hedge.

*(2 marks)*

5 Use the data in Table 6.4 to explain the difference between the figures in your answer to Question 4.

*(2 marks)*

6   Put yourself in the position of a conservation officer talking to a
    farmer who wants to remove a well-kept hawthorn hedge between
    two fields.

    (a)  Summarise the arguments the farmer may put forward for
         removing this hedge.

    *(2 marks)*

    (b)  Use the information in this assignment to outline the argument
         you would put forward for leaving this hedge.

    *(4 marks)*

# Examination questions

1   Read the following passage.

It is estimated that a thousand million tonnes of combined nitrogen
are made available to the biosphere each year from the fixation of
atmospheric dinitrogen by prokaryotic organisms.

 The most effective nitrogen-fixing organisms are those which form
5 symbiotic (mutualistic) relationships with higher plants and the most
important of these relationships is the symbiosis between plants of the
legume family and several species of the bacterial genus *Rhizobium*.
This symbiosis is obviously a true one and it is also highly specialised.
Rhizobia live freely in the soil where they do not fix nitrogen, and are
10 attracted to roots of young legumes by secretions from these roots. The
rhizobia synthesise compounds which deform the root-hair cells of the
host allowing the bacteria to penetrate them. Following infection, the
root-hair cell wall turns inwards and extends into a tube-like 'infection
thread' which penetrates the walls of the cells of the cortex, allowing
15 the bacteria to migrate into the inner cortex of the root. They are finally
released into the cytoplasm of these cells where they multiply and
enlarge into swollen, irregularly shaped bacteroids. Groups of four to
six of these become surrounded by membranes produced by the cell to
form a number of isolated nitrogen-fixing colonies within the cell. Just
20 before the bacteria are released from the infection threads, the cortex
cells of the host in the region of infection multiply rapidly to form the
root nodule.

 In its final structure, the root nodule consists of a central region
containing the nitrogen-fixing bacteroids surrounded by a rhizobia-free
25 area into which the vascular tissue of the root has developed. The
central region has a light pink colour due to the presence of a special
sort of haemoglobin, leghaemoglobin, in the membranes surrounding
each bacteroid group.

The mechanism of nitrogen-fixation is complex and not yet fully
30 understood. The overall equation for the reaction is

$$15ATP + 6H^+ + 6e^- + N_2 \rightarrow 2NH_3 + 15ADP + 15P_i \text{ (inorganic phosphate)}$$

This reaction in the bacteroids is catalysed by the enzyme nitrogenase and makes heavy demands on the photosynthetic product of the host. The enzyme is sensitive to the presence of oxygen and will only fix 35 nitrogen in an anaerobic environment. Although ammonia is the first product, it is rapidly converted into nitrogenous compounds and exported by the nodule through the xylem connected to the host.

(Adapted from *Plants and Nitrogen*, Lewis)

Using information in the passage and your own knowledge, answer the following questions.

(a) What is the meaning of the following terms as used in the passage:

    (i) atmospheric dinitrogen (line 3);

*(1 mark)*

    (ii) combined nitrogen (line 1)?

*(1 mark)*

(b) Draw a fully labelled diagram to show the structure of one of the cortex cells containing bacteroids.

*(3 marks)*

(c) Explain why this symbiosis is described as 'obviously a true one' (line 8).

*(2 marks)*

(d) (i) What is the precise source of the nitrogen fixed by the bacteroids?

*(1 mark)*

    (ii) By what process does the nitrogen reach the bacteroids?

*(1 mark)*

(e) Suggest an explanation for the presence and distribution of leghaemoglobin.

*(4 marks)*

(f) (i) Explain the connection between the nitrogen-fixing reaction and the 'photosynthetic product of the host' (line 33).

*(3 marks)*

    (ii) Why are the demands on the photosynthetic product described as 'heavy' (line 33)?

*(1 mark)*

(g) Describe **one** pathway by which the nitrogen fixed in the nodule might become available for uptake by plants of other species.

*(4 marks)*

(h) Naturally occurring vegetation on a soil with high nitrate content was compared with similar vegetation on a soil with low nitrogen content. The proportion of leguminous plants was much lower on the high nitrate soil. Suggest a reason for this observation.

*(3 marks)*

2 The graph shows how some of the substances found in leaves may be broken down by the action of microorganisms.

percentage of
substance
remaining

— lignin
-- cellulose
--- pectin

time after leaves fell / weeks

(a) (i) Use the graph to calculate the rates of breakdown of lignin, cellulose and pectin between weeks 4 and 12.

*(2 marks)*

(ii) Suggest why leaves are generally broken down much faster than stems.

*(2 marks)*

(b) Describe how the carbon in these substances is made available to growing plants.

*(2 marks)*

3 The diagram shows a nitrogen cycle for an arable farm in which all parts of the crop above ground are removed at harvest.

The figures are in kg of nitrogen ha$^{-1}$ year$^{-1}$.

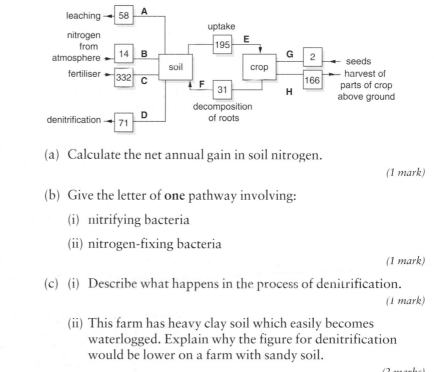

(a) Calculate the net annual gain in soil nitrogen.

*(1 mark)*

(b) Give the letter of **one** pathway involving:

(i) nitrifying bacteria

(ii) nitrogen-fixing bacteria

*(1 mark)*

(c) (i) Describe what happens in the process of denitrification.

*(1 mark)*

(ii) This farm has heavy clay soil which easily becomes waterlogged. Explain why the figure for denitrification would be lower on a farm with sandy soil.

*(2 marks)*

# Reproduction and Development

## Antibodies that stop fertilisation

Many couples delay having children, often because they wish first to establish their careers or create a comfortable home. However, 1 in 7 young couples in the UK do not have children because they are infertile. This means that they are unable to conceive despite regular, unprotected, sexual intercourse for 2 years.

In approximately 30% of infertile couples, the cause is identified only in the woman and in 30% the cause is identified only in the man. In a further 30% of couples, causes can be detected in both partners. Only in about 10% of cases can the underlying cause of infertility not be found using current diagnostic techniques.

Immunological infertility is a rare condition affecting only about 2% of infertile couples. It occurs as a result of antibodies that attach to protein receptors on the surface membranes of sperm cells. As a result, the sperm cells are incapacitated. Antibodies against sperm cells can be present in both men and women. Anti-sperm antibodies in the semen of a man make his sperm cells stick together, so they cannot swim. Anti-sperm antibodies in the cervical mucus of a woman prevent the passage of her partner's sperm to her egg cells. In either case, the sperm cells are incapable of fertilising the woman's eggs naturally.

**Figure 7.1**
These parents are lucky – about 1 in 7 couples in the UK are infertile.

Affected men can be treated with drugs, including antibiotics and steroids, to reduce their production of anti-sperm antibodies. Alternatively, the man's sperm can be collected, washed to remove his anti-sperm antibodies and then introduced into his partner's uterus. If the woman produces anti-sperm antibodies, treatment usually involves collecting her eggs as well as her partner's sperm, washing them to remove antibodies and then mixing them together. You will learn more about this latter technique, called *in vitro* fertilisation (or IVF) later in this chapter.

A biotechnology firm, based in San Diego, hopes to use anti-sperm antibodies as a new method of contraception. Scientists at the firm have isolated the human genes that trigger the production of anti-sperm antibodies. Using genetic engineering techniques, they have spliced these genes into chromosomes in maize plants. When grown in glasshouses, these transgenic maize plants successfully produced anti-sperm antibodies. The company plans to start clinical trials using a plant-based jelly containing the anti-sperm antibodies produced by maize plants.

## Male reproductive system

The human male reproductive system is shown in Figure 7.2.

**Sperm** are produced in the **testes,** which also produce male sex hormones. In humans the testes are enclosed in the **scrotal sac,** an extension of the abdominal cavity. Their semi-external position and a heat exchange system in the arteries and veins leading to and from the testes enables them to be kept at a temperature about 5 °C **lower** than the core body temperature; this is essential for sperm production. The temperature of the testes is fine-tuned by muscles in the scrotal sac. The muscles contract in cold weather, pulling the testes up against the warmer abdomen, and relax in warm weather allowing the testes to drop below the abdomen.

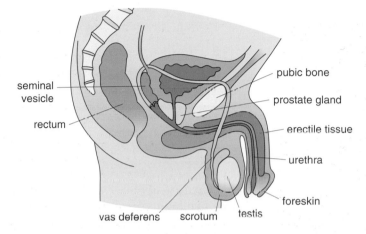

**Figure 7.2**
The human male reproductive organs

**Q** 1 **Why is it thought that tight trousers reduce a man's fertility?**

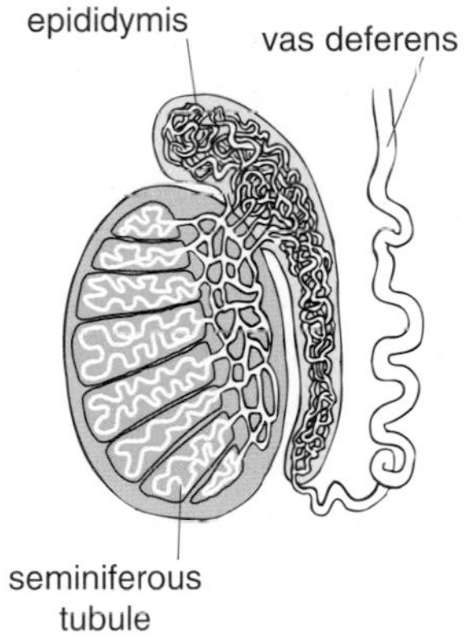

Each testis is subdivided internally into about 200 compartments. Each compartment contains **seminiferous tubules** in which sperm are produced. Each tubule is about 80 cm long, so their combined length is about 500 metres! The seminiferous tubules unite to form a network which leads to the **epididymis**. This is a convoluted tube about 6 metres long where sperm complete their development and are stored (Figure 7.3).

The epididymis leads into a muscular tube, the **vas deferens**. The two vasa deferentia join near the base of the bladder to form the **urethra**, which carries fluids from both the urinary and reproductive systems. Sperm are carried along the urethra in a fluid called **semen**.

Semen from the urethra is ejaculated into the vagina via the **penis**. The penis contains erectile tissue which has spongy areas and blood spaces. During sexual excitement the pressure in the erectile tissue rises because the arteries supplying it dilate and the veins draining it constrict. As a result the penis becomes stiff and erect enabling it to enter the vagina during intercourse.

## Male gamete – sperm

Spermatozoa, or sperm (Figure 7.4), are the smallest mammalian cells.

Each sperm consists of three parts:

● **Head** – This is mainly the **haploid** nucleus, but also has a sac at its tip called the **acrosome**. This is a greatly enlarged lysosome and contains digestive enzymes used to enter the egg.

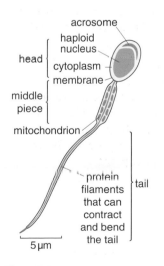

**Figure 7.3**
Internal structure of a testis

**Figure 7.4**
Structure of a human sperm

● **Middle piece** – This contains mitochondria which produce ATP needed for the movement of the tail.

● **Tail** – This is a large flagellum used for moving the sperm.

## Formation of gametes

The formation of gametes is called **gametogenesis**. The overall sequence of events is essentially the same in both sexes:

1  **Multiplication** – diploid cells divide repeatedly by mitosis

2  **Growth** – the daughter cells from the mitotic divisions increase in size

3  **Meiosis** – the products of the growth phase divide by meiosis producing haploid cells

4  **Maturation** – the haploid daughter cells differentiate into gametes: eggs or sperm.

## Spermatogenesis – or sperm production

At puberty diploid cells making up the outer layer (epithelium) of the seminiferous tubule begin to multiply by mitosis. This is followed by a phase of growth. Each cell then undergoes meiosis to form haploid cells. These cells develop into sperm by shedding most of their cytoplasm and reorganising the rest into the middle piece and tail. The entire process of sperm production takes an average of about 64 days. From puberty onwards sperm are produced at a rate of approximately 300 million per day! Figure 7.5 is a summary of this process.

**Figure 7.5**
The process of spermatogenesis. Sperm cells develop as they pass from the outer wall of the seminiferous tubules to the lumen

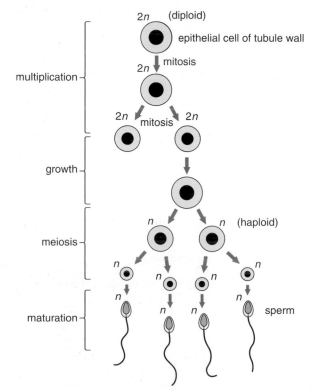

## Microscopic structure of the testes

A testis is packed full of seminiferous tubules and spermatogenesis takes place in their walls (Figure 7.7).

**Figure 7.6**
(a) Group of three seminiferous tubules
(b) Enlarged portion of the wall of a seminiferous tubule showing stages in sperm development

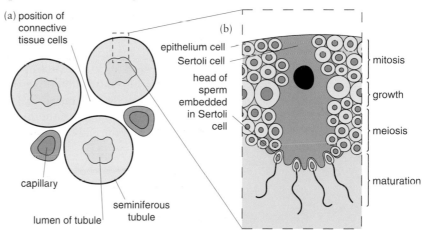

Cell multiplication takes place in the outer part of the wall just beneath the surrounding connective tissue. As cell divisions proceed the daughter cells get pushed towards the lumen of the tubule. (It is here that cells undergoing meiosis can be seen.) The final transformation into sperm takes place in the part of the wall immediately next to the lumen. At this stage the heads of the developing sperm are embedded in large **Sertoli cells** and their tails project into the fluid-filled lumen of the tubule (Figure 7.6).

Since the sperms are genetically different from body cells, they would be in danger from their own immune system were it not for Sertoli cells which act as 'chaperones'. Nutrients and other substances can only reach the developing sperms by passing through the Sertoli cells. When sperm are shed into the seminiferous tubules they are not yet fully motile and are carried passively to the epididymis where they acquire the ability to swim. Even then, a sperm cannot immediately fertilise an egg until it has spent some time in the female reproductive tract. Sperm have a limited life span; if they are not discharged from the male genital tract during intercourse, they degenerate and are either reabsorbed or can be lost via the urine.

**Figure 7.7**
Section of a human testis as seen with a light microscope

## Female reproductive system

The female reproductive system is shown in Figure 7.8.

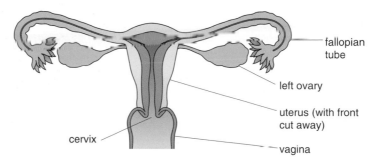

**Figure 7.8**
The human female reproductive system

The **ovaries**, which lie in the abdominal cavity, are about the size and shape of an almond. Like the testes they have a dual function, producing both gametes and sex hormones.

Close to each ovary is the funnel-shaped opening of a **fallopian tube**, in which fertilisation occurs. The walls of the fallopian tube are lined with ciliated epithelia and contain smooth muscle. A combination of the movement of the cilia and peristaltic muscle contractions serves to transport eggs down to the **uterus**. The wall of the uterus has two distinct layers reflecting its two functions. The bulk of the uterus wall consists of smooth muscle, the **myometrium**, which functions to expel the fetus at birth. The inner lining, or **endometrium**, is concerned with the anchorage and nutrition of the embryo. The neck of the uterus, or cervix, opens into the **vagina**.

## Female gamete or egg

**Figure 7.9**
The structure of an egg

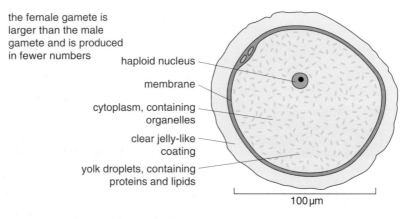

the female gamete is larger than the male gamete and is produced in fewer numbers

haploid nucleus
membrane
cytoplasm, containing organelles
clear jelly-like coating
yolk droplets, containing proteins and lipids

100 µm

Since an egg does not have to propel itself, its structure is more simple than that of a sperm. On the other hand it contains nutrients to help sustain it through the earliest stages of its development. It is therefore much larger than a sperm. A human egg is 0.1 mm (100 µm) in diameter, compared with a sperm whose head is only 2.5 µm across. The egg has a large haploid nucleus surrounded by cytoplasm, which contains the usual organelles and enzymes found in all body cells. Outside the plasma membrane is a glycoprotein coat which, because of its jelly-like consistency, is called the jelly coat.

### Oogenesis – or egg production

As in spermatogenesis, egg production involves four phases in which the cells go through multiplication, growth, meiosis and maturation (Figure 7.10).

In a female the development of the gamete begins before birth. After repeated divisions by mitosis the products enter a phase of growth, increasing in size. On completion of its growth it now enters prophase of the first division of meiosis, but development is then suspended until after puberty.

**Figure 7.10**
The process of oogenesis begins in the ovaries of a developing fetus but it is not completed unless the egg is fertilised

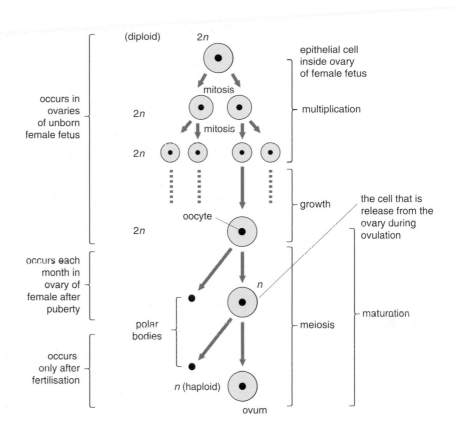

There are two important differences between spermatogenesis and oogenesis.

- Egg production begins before birth but is not completed until fertilisation, which could be 50 years later.

- The divisions accompanying meiosis are very unequal. The result is that only one functional ovum is produced, the other products being tiny functionless **polar bodies**.

## Microscopic structure of the ovary

Figure 7.11 shows an ovary.

The ovary consists of a mass of tissue in which there are many blood vessels. These blood vessels not only supply the organ with the oxygen and nutrients it needs and remove waste products such as carbon dioxide, but also transport the hormones that are responsible for controlling the process of reproduction. Within the ovary are **follicles** at different stages of development. Each follicle consists of a hollow ball of cells, the follicle cells, surrounding a developing egg cell or **oocyte**.

At the time of birth there are about one million primary follicles in each ovary. Like the oocytes within them the follicles remain in suspended development until puberty, some twelve years later. At this time all but about 400 000 have degenerated, of these only 400 will be released into the oviduct during the reproductive life of a woman. At monthly intervals about 20–25 follicles begin to develop further, from these only a single oocyte is released.

**Figure 7.11**
Internal structure of the ovary showing stages in follicle development in a human ovary as it appears under the microscope

We can recognise three main stages in development (Figure 7.12):

- **The follicular stage** – During this first part of the cycle one or more follicles start to develop. The follicle cells surrounding the oocyte grow and divide becoming several layers thick. The follicle eventually reaches a diameter of about 15 mm. This is called a mature **ovarian follicle**.

  The follicular stage leads to the development of a mature gamete. The follicle produces hormones that affect other parts of the reproductive system and prepares it for a possible pregnancy.

- **Ovulation** – The oocyte is released from the ovary still surrounded by a layer of follicle cells and passes down the fallopian tube towards the uterus.

- **The luteal phase** – Most of the follicle cells remain in the ovary after ovulation. They continue to develop and form a structure called the **corpus luteum**. The corpus luteum produces hormones. These hormones keep the reproductive system in a condition that will allow the implantation of a fertilised egg and the maintenance of pregnancy. If there is no fertilisation, the corpus luteum degenerates.

If pregnancy takes place, this cycle is inhibited. If pregnancy does not take place the cycle begins again immediately.

**Figure 7.12**
The internal structures of an ovary showing stages in follicle development

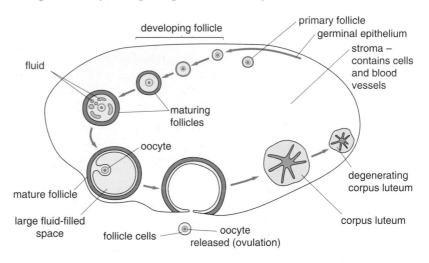

---

**Extension box 1**

## Hormones controlling the female reproductive cycle

### Follicular stage

The pituitary gland is situated underneath the brain. At the start of the cycle this gland secretes a hormone called **follicle-stimulating hormone (FSH)**. FSH travels round the body in the blood (1). In the ovary it causes the follicle cells surrounding one or more of the future egg cells to begin to divide, with the result that the follicle becomes a mature ovarian follicle.

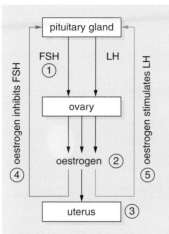

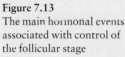

**Figure 7.13**
The main hormonal events associated with control of the follicular stage

The developing follicle cells secrete the hormone **oestrogen** (2) which:

- stimulates the growth of the **endometrium** and its blood supply (3)
- inhibits further secretions of FSH by the pituitary gland (4)
- stimulates the pituitary gland to secrete a second hormone, **luteinising hormone** (**LH**) (5).

These events are summarised in Figure 7.13.

## Ovulation

As the follicular stage progresses, the developing follicle steadily increases in size and finally becomes a mature ovarian follicle. As it gets larger it produces more and more oestrogen, with the result that the oestrogen concentration in the blood increases. Most of the LH produced during the follicular stage of the cycle is stored in the pituitary gland. When the concentration of oestrogen in the blood is high enough the release of LH is triggered so that there is a rapid increase in the concentration of LH in the blood. It is this surge of LH that brings about ovulation. The mature ovarian follicle bursts and the oocyte, together with some of the follicle cells that surround it, leaves the ovary and passes into the fallopian tube on its way to the uterus. It is at this stage, with the oocyte in the fallopian tube, that fertilisation can take place.

## Luteal stage

The surge of LH that brings about ovulation also has an effect on the follicle cells that remain in the ovary (1). The follicle now becomes a structure known as the **corpus luteum**. The corpus luteum secretes some oestrogen and a large amount of another hormone, **progesterone** (2).

- Progesterone continues to stimulate the growth of the endometrium and its blood supply, and also stimulates the development of nutrient fluid glands in this lining layer (3).

- The high concentrations of oestrogen and progesterone inhibit the production of FSH and LH (4). Without these hormones the cells of the corpus luteum start to shrivel and produce less and less oestrogen and progesterone. Since these are the hormones that inhibit the production of FSH, the cycle can start again.

These events are summarised in Figure 7.14.

In humans, at the end of the luteal stage, the levels of oestrogen and progesterone fall. As a result, spiral blood vessels in the lining of the human uterus contract and cut off the blood supply to the cells that form the uterus lining, causing them to die. These spiral vessels then dilate again. The increased blood flow causes the uterus lining to be shed and lost from the body, an event known as **menstruation**. Because of menstruation we often refer to these phases as the **menstrual cycle**. Figure 7.15 shows the hormonal events and changes that take place in the endometrium during one complete menstrual cycle.

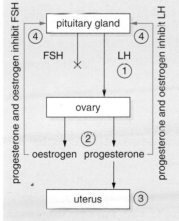

**Figure 7.14**
The main hormonal events associated with control of the luteal stage

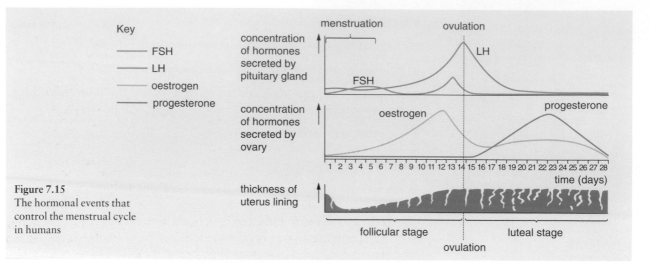

**Figure 7.15**
The hormonal events that control the menstrual cycle in humans

## Fertilisation

On entering the vagina, the semen initially coagulates but soon turns to liquid again enabling the sperm to swim freely. The acidic pH in the vagina is hostile to sperm but is partially neutralised by the alkalinity of the semen. The 200 million or so sperm in a typical ejaculate are deposited just outside the cervix of the uterus about 10–15 cm from the egg.

The egg is fertilised in the fallopian tube. This occurs up to 18 hours after ovulation.

## Movement of sperm in the female genital tract

Ejaculation usually occurs with sufficient force to propel the sperm to the upper end of the vagina. They must now get from here to the upper part of the fallopian tube. By rhythmic wave-like movements of their tails they swim through the mucus in the cervix. At the time of ovulation this mucus is thin and watery as the glycoprotein chains within it run parallel with each other. This makes it easier for the sperm to swim through it. Though sperm can survive in the female tract for up to 48 hours, the average time taken to reach the oviduct is 4–6 hours. It is known that sperm are able to travel the distance to the oviduct within 5 minutes of intercourse, so they must be helped in some more active way. This is now thought to be the active muscular contractions of the uterus during intercourse.

During the journey through the female reproductive tract, the sperm undergo a process called **capacitation**, in which they acquire the ability to fertilise the oocyte. Capacitation takes about 5–7 hours and is thought to involve the removal of certain proteins from the membrane overlying the acrosome. Usually no more than a few hundred sperm reach the area surrounding the oocyte and only one fertilises it. Contact between the sperm and oocyte seems to be a matter of chance. Before the sperm can effect fertilisation, the **acrosome reaction** must take place (Figure 7.16).

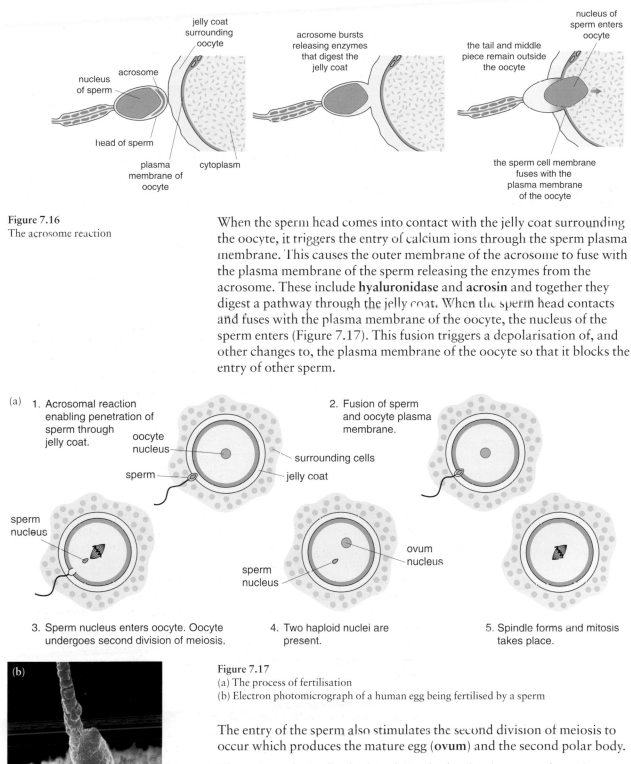

**Figure 7.16**
The acrosome reaction

When the sperm head comes into contact with the jelly coat surrounding the oocyte, it triggers the entry of calcium ions through the sperm plasma membrane. This causes the outer membrane of the acrosome to fuse with the plasma membrane of the sperm releasing the enzymes from the acrosome. These include **hyaluronidase** and **acrosin** and together they digest a pathway through the jelly coat. When the sperm head contacts and fuses with the plasma membrane of the oocyte, the nucleus of the sperm enters (Figure 7.17). This fusion triggers a depolarisation of, and other changes to, the plasma membrane of the oocyte so that it blocks the entry of other sperm.

**Figure 7.17**
(a) The process of fertilisation
(b) Electron photomicrograph of a human egg being fertilised by a sperm

The entry of the sperm also stimulates the second division of meiosis to occur which produces the mature egg (**ovum**) and the second polar body.

There are now two haploid nuclei in the fertilised egg, one from the sperm and the other from the ovum. Several hours after the sperm has penetrated the oocyte, the nuclear envelopes of these nuclei break down, a spindle forms and mitosis occurs.

## From fertilisation to implantation

The **zygote**, the first cell formed as the result of fertilisation, is slowly carried down the fallopian tube by the action of the cilia, possibly aided by muscular action of the walls. Here it divides by mitosis into two cells, then four and so on, forming a solid ball of cells (Figure 7.18).

**Figure 7.18**
As the fertilised egg moves down the fallopian tube it divides into many cells forming a blastocyst

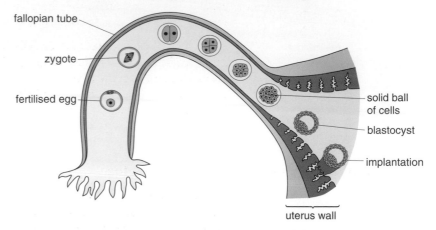

By the time it has reached the uterus, about 4 days after fertilisation, the little ball of cells has developed a cavity and is now called a **blastocyst**. This consists of an inner cell mass from which the embryo eventually develops and an outer layer of cells, the **trophoblast**, which is concerned with the future nourishment of the embryo.

## Implantation

By the time the blastocyst arrives in the uterus it consists of about 100 cells. It remains free in the uterus for 2–3 days, continuing to develop and nourished by secretions from the uterus. When the blastocyst comes into contact with the endometrium, it gradually becomes buried within this tissue. This occurs over a period of about a week (Figure 7.19).

**Figure 7.19**
Implantation of the blastocyst into the uterine lining

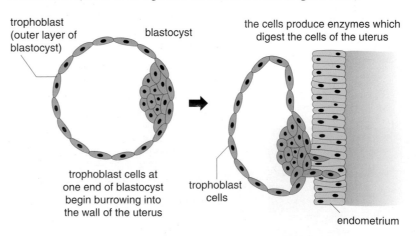

The outer layer of the blastocyst forms branched extensions, or villi, which secrete enzymes that digest the superficial tissues of the endometrium. The nutrients released are used by the embryo and for a week after implantation these products of digestion are the only source of food for the embryo. As the digestion extends to the walls of the maternal blood vessels, the villi become surrounded by maternal blood spaces. Since the villi are finely divided they provide a large surface area over which the exchange of gases and nutrients can occur. Later these villi are replaced by larger and more extensive connections between mother and fetus. This structure is known as the **placenta**.

The trophoblast has a second, equally important function; it secretes the hormone **human chorionic gonadotrophin** (**hCG**). This maintains the corpus luteum which will continue to produce oestrogen and progesterone. The production of hCG by the blastocyst can be detected in urine a day or so before implantation and forms the basis for a pregnancy test.

| | |
|---|---|
| **Extension box 2** | ## Treating infertility |

### Treating infertility

In some women the pituitary gland fails to produce enough FSH, with the result that mature ovarian follicles do not develop in the ovary and ovulation does not occur. This can be remedied by injections of FSH or a synthetic equivalent, the so-called **fertility drug**. So effective is this treatment that it may cause several eggs to be produced at the same time, resulting in twins or even greater multiples.

Some women fail to ovulate because they secrete too much oestrogen which has the effect of inhibiting FSH secretion by the pituitary gland. They can be treated with non-steroidal drugs such as clomiphene which opposes the action of oestrogen, thus releasing the pituitary gland from its inhibition.

### In vitro fertilisation

In vitro fertilisation (IVF) is commonly known as the test tube baby technique. The first successful use of the technique came with the birth of Louise Brown in 1978. It was developed as a treatment for certain forms of infertility, mainly blocked or damaged oviducts. The term in vitro literally means 'in glass' and refers to the fact that fertilisation takes place outside the body, usually in a plastic dish. The technique may also be used if it is desirable to screen embryos for genetic defects before allowing pregnancy to proceed. The main steps involved in the procedure are as follows:

1 A woman being treated is injected with a hormone such as FSH which stimulates the growth of many follicles and therefore eggs.

2 Just before ovulation 'ripe' eggs are collected.

3 The male partner produces a sperm sample.

4 The eggs are then matured for up to 24 hours.

5 When mature, the sperm are added to the eggs and left for fertilisation to occur. This process can be assisted by micro-injecting the nucleus of the sperm into the egg.

6  Two or three days later any embryos that have formed are examined.

7  A maximum of three are then transferred into the uterus.

This technique raises a number of important ethical issues. Probably the most important is the fate of unused embryos. These can be frozen and stored for future use, disposed of, or used for research.

## Male infertility

In men, infertility is sometimes caused by the semen containing too few sperm (low sperm count). This may be remedied by treatment with natural or synthetic androgens such as testosterone. Another problem that some men face, particularly as they get older, is failure of the penis to become erect. The drug **Viagra** (sildenafil), which can be taken orally, rectifies this condition. It is an enzyme inhibitor and indirectly causes the smooth muscle surrounding the main cavities in the penis to relax so that more blood can be pumped into them during the erection process.

## Development of an embryo

The period of development, from fertilisation to birth, is called **gestation**. During this time the new individual increases in size by more than a thousand million times, a relative growth rate that is far in excess of growth after birth. Development also involves a great increase in complexity. The relatively uniform cells of the inner cell mass of the blastocyst eventually differentiate into more than 200 kinds of cells which are organised into tissues and organs.

By the time it is a month old the embryo has the beginnings of a gut, kidneys and brain, and a heart which is already beating. By two months, all the main organ systems are present and the embryo is called a **fetus**.

**Figure 7.20**
A temperature chart to use with question 2.

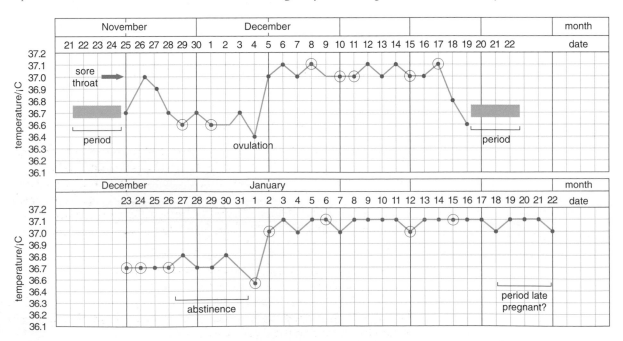

**Q** 2 Many women who are trying to become pregnant are asked to keep a record of their daily temperature. Also noted are the occasions when sexual intercourse took place and any illness. Study the temperature chart (Figure 7.20) and answer the following questions.
(a) Why is it important to take the temperature at the same time every day?
(b) Why would the woman be advised to record any illness?
(c) What is the normal temperature: before ovulation, after ovulation?
(d) What effect does pregnancy have on temperature?
(e) On what date(s) is she most likely to have conceived?

## The placenta

At the end of gestation or pregnancy the human placenta is discarded, but it is essential for the first nine months of life.

When the small ball of cells, the blastocyst, embeds itself in the wall of the uterus, the placenta is the first organ to develop. In early pregnancy it grows much more rapidly than the embryo (Figure 7.21).

**Figure 7.21**
Growth curves showing the increase in mass of the placenta and fetus during pregnancy. Initially the placenta has a greater mass, though both are very small

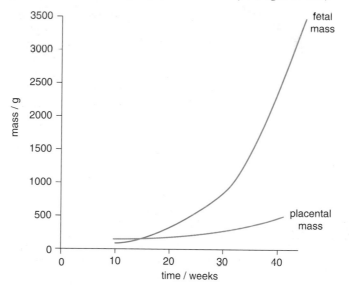

### Structure of the placenta

The placenta is an organ formed partly from membranes surrounding the developing fetus and partly from the lining of the uterus, the endometrium. When fully developed it is approximately 20 cm across and 3 cm thick. It is connected to the fetus by the **umbilical cord** which contains **two umbilical arteries** carrying blood from the fetus to the placenta, and an **umbilical vein** carrying blood from the placenta to the fetus.

In the placenta, the blood of the mother and the fetus come very close together but do not mix.

**Q** 3 Give two reasons why mixing the blood of the mother and fetus would be harmful to the fetus.

As the placenta develops, many tree-like villi grow into the endometrium, digesting the walls of the maternal blood vessels. As a result villi are surrounded by a very thin, continuous 'pool' of maternal blood, supplied by the **uterine arteries** and drained by the **uterine vein** (Figure 7.22). Each villus contains a network of fetal capillaries.

**Figure 7.22**
Part of the placenta showing maternal and fetal blood supply

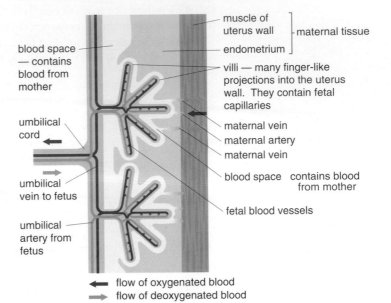

## Function of the placenta

The placenta has three main functions.

- It allows substances to be exchanged between the blood of the fetus and the blood of the mother.

- It is important in producing hormones necessary for the maintenance of pregnancy: hCG, oestrogen and progesterone.

- It allows antibodies to be transferred from the mother to the fetus, so that for some weeks after birth the infant has immunity to the same diseases as its mother.

Oxygen and substances such as glucose, amino acids, mineral salts and water pass from the mother's blood into the fetal blood. Waste products such as urea and carbon dioxide are removed from the blood of the fetus to that of the mother. Substances with small molecules such as oxygen, carbon dioxide and urea cross between fetal and maternal blood by simple diffusion; glucose crosses by facilitated diffusion; and amino acids cross by active transport. Maternal antibodies are taken into the villi by pinocytosis.

Exchange of oxygen between the two blood systems is made possible by the fact that fetal haemoglobin has a greater affinity for oxygen than adult haemoglobin (see Chapter 10).

The epithelium lining the placental villus bears microvilli, which increase its surface area. It also contains numerous mitochondria which provide energy for active transport of amino acids (Figure 7.23).

**Figure 7.23**
Transport across the placenta:
(a) transport pathway
(b) mechanisms by which
different substances move
across this pathway

In the fully developed human placenta, only three fetal layers separate fetal and maternal blood: the capillary endothelium, a thin layer of connective tissue and the epithelium covering the villi. The barrier separating fetal and maternal blood is therefore very thin, so the diffusion path between fetal and maternal blood is very short: about 3.5 μm.

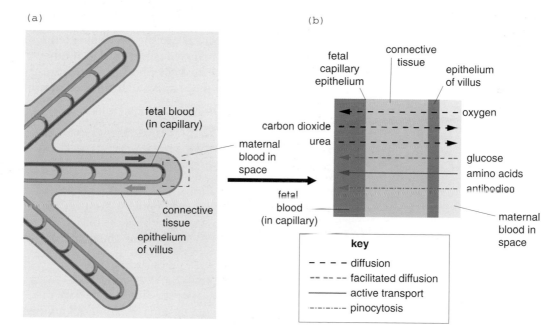

Like all exchange organs, the placenta has a number of adaptations which ensure efficient exchange.

**Large surface area**

- The exchange surface of the placenta is folded into villi
- These villi have microvilli on their plasma membrane.

**Maintaining a maximum difference in concentration on either side of the exchange surface**

- Continuous supply of oxygenated maternal blood in the uterine arteries
- Removal of oxygenated fetal blood by the umbilical vein
- Fetal haemoglobin has a much higher affinity for oxygen than adult haemoglobin
- The blood of the mother flowing in the opposite direction to that of the fetus, forming a counter-current system.

**Short diffusion path**

- There is a reduction in the number of cell layers between the blood supply of the mother and that of the fetus, due to breakdown of maternal blood vessels.

**Q** 4 The fetus gets its oxygen by diffusion from its mother's blood. Use Fick's Law to explain how the placenta ensures rapid diffusion of oxygen.

**Q** 5 In what ways is the composition of the blood in the umbilical vein:
(a) similar to that in the hepatic portal vein
(b) different from that in the hepatic portal vein?

**Q** 6 The body temperature of the developing fetus is approximately 0.5 °C higher than the body temperature of its mother.
(a) Explain what causes the higher body temperature.
(b) Describe the role of the placenta in helping to control fetal body temperature.

## Maternal changes in pregnancy

As the fetus develops, changes are also occurring in the mother. The smooth muscle fibres of the uterus wall enlarge greatly and the uterus grows from its normal mass of about 50 g to over 1 kg by the time the baby is born. In response to oestrogen, the ducts of the breast tissue enlarge, whilst progesterone stimulates the growth of the secreting tissue in the breasts.

The mother's body responds to being pregnant in various ways. For example her body mass, appetite, thirst, metabolic rate, ventilation rate, cardiac output, blood volume and red blood cell number all increase; as does the release of calcium ions and glucose into the bloodstream. Her thermal balance changes too and her food requirements increase, for she is in effect feeding two people. Her diet must include plenty of carbohydrate (for energy), protein (for growth), iron (for haemoglobin), calcium (for bones) and various vitamins.

These adaptations generally minimise the stresses imposed on her body and provide the best environment for the growing fetus. Let us look at a few of these in more detail.

### Cardiovascular system

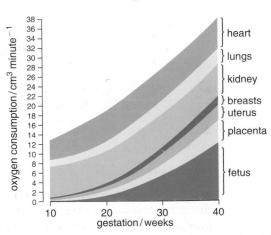

Figure 7.24
Increase in oxygen consumption during pregnancy. The main part of the increase goes to the fetus

The increased load on the heart in pregnancy is due to greater need for oxygen in the tissues.

- The fetus's body and organs grow rapidly and its tissues have an even higher oxygen consumption than the mother's.

- The growth of most maternal tissues, not just the breast and uterus, increases oxygen requirements.

- The mother's muscles have to work harder to move her increased size and that of the fetus.

- Cardiac output is the product of stroke volume and heart rate. It is greater during pregnancy because the heart beats faster combined with a small increase in stroke volume. There is an increase in cardiac muscle so that the heart chambers enlarge and output increases by 40% (Figure 7.25).

**Figure 7.25**
Cardiac output in pregnancy. The increase occurs very early, reaching its maximum at 20 weeks.

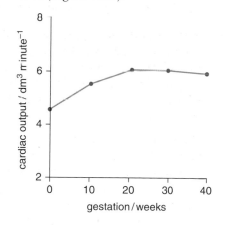

## Blood volume

Maternal blood volume increases. The changes in the volume of the plasma being greater than the increase in the number of red blood cells.

**Figure 7.26**
Changes in blood volume, plasma volume and red cell mass during pregnancy, when compared with non-pregnant values

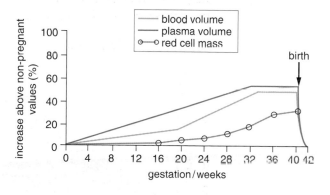

## Thermal balance

The fetus produces heat during respiration. Due to its rapid growth, the fetus's respiration rate is higher and consequently so is its heat production. A fetus must be able to remove this excess heat.

The mother's blood is cooler and thus there is a heat gradient. As this comes close to the fetal blood in the placenta, so heat is transferred to the mother's blood. The mother can then lose this excess heat from her own body.

## Birth

A normal pregnancy lasts about 38 weeks from implantation, 40 weeks from the last period. During the last few weeks the fetus usually comes to lie with its head down against the cervix. The process of childbirth, or labour, involves three stages.

1 The first stage is the longest, lasting up to twenty-four hours. The function of this stage is to stretch the cervix. The uterus begins to undergo spontaneous weak contractions beginning approximately every 30 minutes and increasing in strength and frequency until they are coming every 3 minutes. These contractions are caused by the hormone **oxytocin**, secreted by the **posterior pituitary gland**. When the cervix is fully dilated, the second stage begins.

2 This stage lasts only minutes and results in the expulsion of the baby aided by voluntary contractions of the mother's abdominal muscles. At this point the umbilical cord shuts down, isolating the baby from the mother. The resulting rise in carbon dioxide content of the blood stimulates the baby's first breath.

3 The final stage consists of the expulsion of the placenta, the afterbirth, and occurs about half an hour after the baby is born.

## Hormonal changes during and after pregnancy

When the oocyte is fertilised and the woman becomes pregnant the corpus luteum persists. This is due to the hormone, **human chorionic gonadotrophin (hCG)**, which is secreted by the blastocyst and then by the developing placenta. It is the basis of pregnancy tests: excess hCG is excreted, and its presence in the urine is detected and used as an indication that the woman is pregnant.

The production of hCG begins soon after fertilisation, and its concentration in the blood stream increases to a peak after about two months followed by a slow decline.

The corpus luteum continues to secrete progesterone which, together with a small but steady secretion of oestrogen, maintains the development of the uterus and, of course, prevents menstruation.

After the first three months of pregnancy the corpus luteum begins to degenerate and the role of secreting oestrogen and progesterone is taken over by the placenta. In this way the endometrium of the uterus is maintained in a suitable state throughout pregnancy. These hormones also inhibit the anterior pituitary from producing FSH, thus preventing further mature ovarian follicles from developing in the ovary (Figure 7.27).

**Figure 7.27**
Changes in hormone
concentrations during
pregnancy

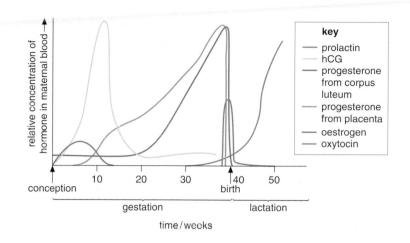

## Hormonal control of birth

Towards the end of pregnancy, the level of oestrogen in the blood rises
while that of progesterone falls. Oestrogen promotes uterine contraction,
whilst progesterone inhibits it. Oestrogen makes the muscle in the uterus
more sensitive to another hormone, oxytocin, which is secreted by the
posterior pituitary gland. This hormone causes the muscles in the uterus
to contract (Figure 7.28).

**Figure 7.28**
Hormonal control of uterine
contractions

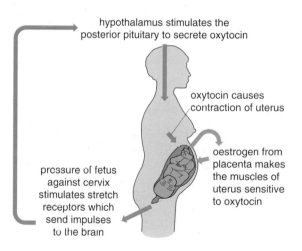

## Changes in circulation

The placenta is the means by which the fetus obtains oxygen. Because of
this the lungs of the fetus are not functional and the fetal circulation is
different from that of the adult.

The basic plan of the fetal circulation is shown in Figure 7.29.

There are a number of important differences between the circulation of
the fetal and adult blood systems.

The umbilical arteries carry deoxygenated blood from the aorta of the
fetus to the placenta where the blood is oxygenated. The umbilical vein
then carries oxygenated blood from the placenta to the vena cava of the

fetus, from where it enters the right atrium of the heart. As the lungs are non-functional, most of the blood bypasses them by flowing through a hole in the heart, the **foramen ovale**, which connects the right and left atria. The foramen ovale is guarded by a flap valve which prevents blood flowing back from the left to the right atrium. Thus most of the oxygenated blood passes through the foramen ovale into the left side of the heart, which pumps blood to the brain, limbs and organs of the body. The rest of the blood, in the right atria, passes to the right ventricle and into the pulmonary artery. In an adult this artery carries blood to the lungs but only a small portion of the fetal blood travels to these organs. The lungs of the fetus, in their unexpanded state, offer a high resistance to flow. An alternative bypass is present, the **ductus arteriosus**, this is a short vessel which connects the pulmonary artery with the aorta. Most of the blood therefore crosses to the aorta via the ductus arteriosus and then travels to the legs, trunk and placenta.

**Figure 7.29**
The circulation of blood in a fetus

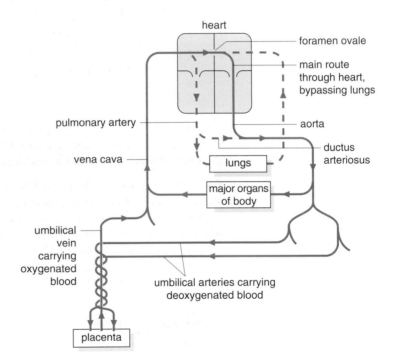

At birth, the placenta is replaced by the lungs as the organ of gas exchange. This means that all blood travelling into the right side of the heart must now flow to the lungs. Within a few minutes of birth, the ductus arteriosus permanently constricts so that all the blood leaving the right ventricle is sent to the lungs. This, coupled with the closure of the umbilical vein, results in the blood pressure in the left atrium exceeding that in the right atrium. This has the effect of closing the valve of the foramen ovale. A few days after birth, the foramen ovale becomes sealed by the fusion of its valve with the atrial wall.

Besides these structural changes, birth also triggers a switch from the production of fetal haemoglobin to adult haemoglobin, though it takes 3–4 months for all the fetal red blood cells to be replaced.

## Hormonal control of lactation

Throughout pregnancy oestrogen and progesterone stimulate the development of milk-producing tissue in the breasts (Figure 7.30).

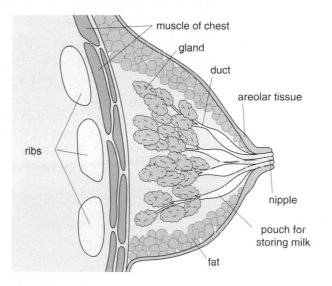

**Figure 7.30**
Structure of human mammary gland

The high concentrations of progesterone also inhibit the hormone **prolactin**. At birth as the progesterone concentration falls, prolactin is no longer inhibited. Prolactin is now able to stimulate the breasts to produce milk.

When the baby suckles (1), nerve impulses travel to the hypothalamus (2), an area in the mother's brain, causing the pituitary gland to produce oxytocin and prolactin (3). The oxytocin stimulates the muscles in the walls of the milk ducts to contract, squeezing milk into the baby's mouth (4) and the prolactin stimulates the production of more milk (5) (Figure 7.31).

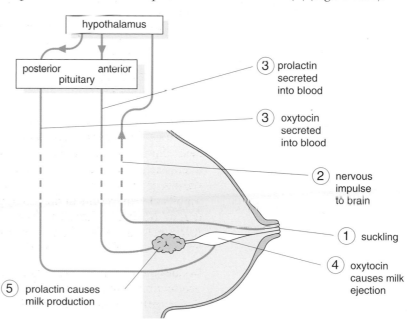

**Figure 7.31**
How suckling promotes the production and ejection of milk

One of the other effects of suckling is that prolactin helps to prevent the secretion of two hormones, FSH and LH. Thus ovulation does not happen. A mother is therefore less likely to conceive as long as she continues to breast feed her baby. Breast milk may also be considered better for the baby because it is free from bacteria and contains antibodies and essential nutrients in the right proportions.

Milk contains the necessary nutrients for the baby and is especially rich in calcium to help its bones grow. However, after 3–4 months the baby will need solid food because milk contains no fibre and practically no iron.

## Summary

- The testis and ovary are the sites of gamete formation (gametogenesis).

- Gametogenesis involves the processes of mitosis, growth, meiosis, and maturation.

- Gametogenesis in the male results in the formation of many small sperm, whereas in the female it results in the formation of a few large ova.

- The structure of a mature sperm cell is adapted so that it can move and the acrosome reaction enables it to penetrate the oocyte membranes.

- The zygote develops into a blastocyst and is implanted into the wall of the uterus.

- The blood system of the developing fetus contains two umbilical arteries taking blood to the placenta and one umbilical vein taking blood from the placenta.

- As the lungs of a fetus are not functional, the foramen ovale, a channel between the right and left atria of the heart and the ductus arteriosus, a vessel linking the pulmonary artery to the aorta, cause most of the blood to bypass the lungs.

- The placenta is adapted for exchange and active transport between the blood systems of the mother and the fetus.

- Pregnancy affects thermal balance and causes changes in blood volume and cardiac output.

- Hormones interrupt the normal menstrual cycle during pregnancy, control birth and lactation.

# Assignment

## Hormones as contraceptives

It is not difficult to learn facts, but they are of little use to you if you do not understand them. Understanding allows you to interpret information and apply your knowledge to new situations. This assignment is based on a passage about the use of hormones as contraceptives. The questions it contains have been designed to test your understanding of underlying biological principles. In addition, the last question involves you bringing together ideas from different parts of the specification. As you continue with your A-level studies, you will appreciate more and more that a study of biology involves linking together individual topics in different ways.

When ovulation has taken place, two steroid hormones are produced by the ovary. These hormones are oestrogen and progesterone. The **combined contraceptive pill** contains a mixture of these hormones. They are both lipids, as the fact that they are steroids suggests, but progesterone is not absorbed very well by the cells lining the gut. As a result of this, a synthetic form of the hormone is used. The combined pill works by inhibiting the release of FSH and LH by the pituitary gland, preventing ovulation from taking place. If the ovaries of women taking these pills are compared with those of women not taking hormonal contraceptives, a number of differences can be seen.

Oestrogen can have side effects, so a **mini-pill** has been produced which contains progesterone only. It works in a different way to the combined pill, as it does not normally inhibit ovulation. Instead, it interferes with the process of meiosis. In rare cases, fertilisation does take place, in which case implantation is prevented. With the mini-pill, FSH is not inhibited, thus oestrogen concentrations follow the same pattern as in a woman not taking oral contraceptives.

A third type of oral contraceptive is known as the **morning-after pill**. One form involves taking a single large dose of progesterone on the day following intercourse. This stimulates the lining of the uterus to develop. As it is a single dose, the concentration of progesterone in the blood then decreases rapidly. Such an effect on the uterus lining prevents implantation from occurring.

A possible future development involves using a synthetic polypeptide, which is very similar to GnRH (gonadotrophin-releasing hormone). This is a hormone produced by the hypothalamus. It triggers the release of FSH and LH by the pituitary gland. GnRH is a small polypeptide containing 10 amino acids. By substituting its constituent monomers, over 300 similar molecules have been produced. Some of these have been found to block GnRH receptor sites on plasma membranes of cells in the pituitary gland. This could represent a major advance in producing a safe, but reliable and cheap, contraceptive.

**Figure 7.32**
There are a number of different types of oral contraceptive, but most of them are based on the hormones that normally control reproduction

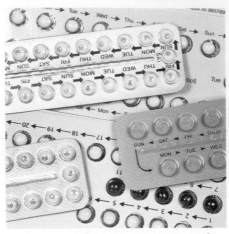

Now answer these questions. In each answer, make use of the information provided, but try to add a little more detail from your own knowledge as well.

1 Three groups of 30-year-old women, who were not pregnant, were each using a different type of contraceptive. Copy the table below. Complete it by using ticks to show the stages in follicle development that might be found in the ovaries of the women in each of the three groups.

| Type of contraceptive used | Stage in follicle development that might be found in ovaries | | |
|---|---|---|---|
| | immature follicle | mature ovarian follicle | corpus luteum |
| Not using hormonal | | | |
| Combined contraceptive pill | | | |
| Mini-pill | | | |

*(3 marks)*

2 Explain what is meant by:

(a) a steroid

(b) a gonadotrophin

*(2 marks)*

3 Use your knowledge of reproduction and the hormones that control it to explain how each of the following normally prevents pregnancy:

(a) the combined contraceptive pill

(b) the mini-pill

(c) the morning after pill

*(6 marks)*

4 Use your knowledge of proteins to explain:

(a) how the 300 similar molecules to GnRH, described in the last paragraph in the passage, differ in structure from GnRH.

*(2 marks)*

(b) how a contraceptive based on one of these molecules could prevent GnRH from stimulating the pituitary gland to produce FSH.

*(3 marks)*

(c) why a contraceptive based on one of these molecules would have to be injected rather than swallowed.

*(2 marks)*

# Examination questions

1 The flowchart summarises the way in which the process of sperm formation is thought to be controlled by hormones.

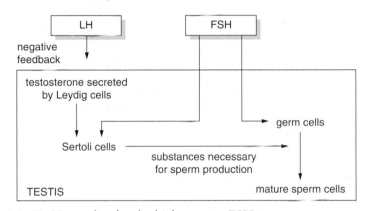

(a) (i) Name the gland which secretes FSH.

*(1 mark)*

(ii) Complete the table to show the percentage of each type of cell which would contain a **Y** chromosome.

*(2 marks)*

| Cells | Percentage of cells containing a Y chromosome |
|---|---|
| Sertoli cells | |
| Mature sperm cells | |

(b) Explain why increasing the amount of LH in the blood leads to an increase in the amount of Golgi apparatus in Leydig cells.

*(2 marks)*

(c) The graph shows changes in LH and testosterone concentrations in the blood of an adult man.

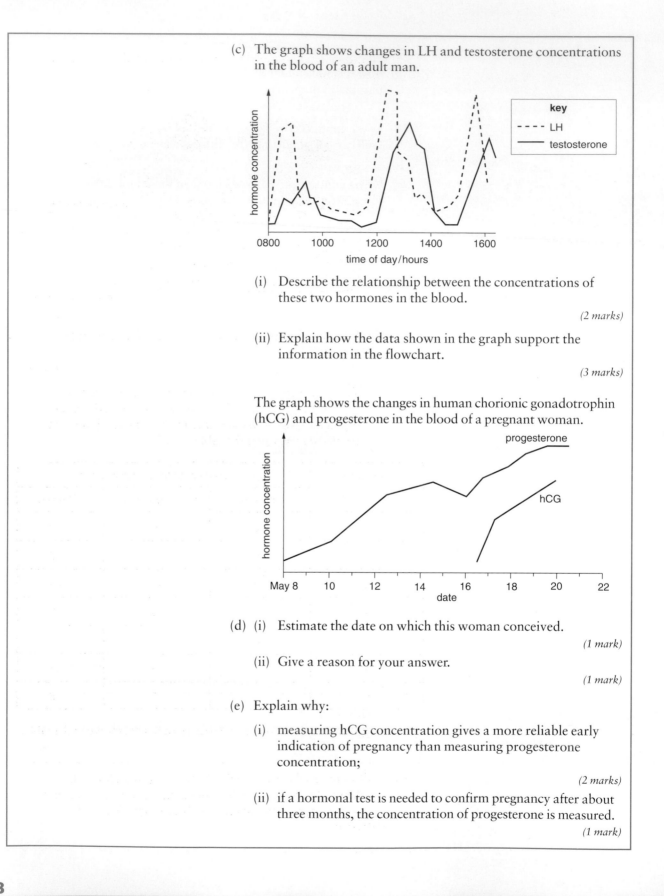

(i) Describe the relationship between the concentrations of these two hormones in the blood.

*(2 marks)*

(ii) Explain how the data shown in the graph support the information in the flowchart.

*(3 marks)*

The graph shows the changes in human chorionic gonadotrophin (hCG) and progesterone in the blood of a pregnant woman.

(d) (i) Estimate the date on which this woman conceived.

*(1 mark)*

(ii) Give a reason for your answer.

*(1 mark)*

(e) Explain why:

(i) measuring hCG concentration gives a more reliable early indication of pregnancy than measuring progesterone concentration;

*(2 marks)*

(ii) if a hormonal test is needed to confirm pregnancy after about three months, the concentration of progesterone is measured.

*(1 mark)*

2 (a) **Figure 1** shows some cells from a mammary gland which is secreting milk.

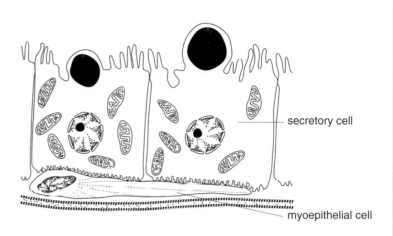

(i) Explain why there are large numbers of mitochondria in these cells.

*(2 marks)*

(ii) The cells were supplied with amino acids which were radioactively labelled. The radioactivity of various cell organelles was then measured over the next 6 hours. The results are shown in the table.

| Time/hours | Radioactivity/arbitrary units | | |
|---|---|---|---|
| | Ribosomes | Golgi apparatus | Vesicles |
| 0 | 0 | 0 | 0 |
| 1 | 84 | 12 | 2 |
| 2 | 76 | 58 | 45 |
| 3 | 31 | 64 | 57 |
| 4 | 8 | 36 | 42 |
| 5 | 2 | 16 | 21 |
| 6 | 6 | 5 | 8 |

Suggest an explanation for the pattern of figures shown in the table.

*(3 marks)*

(b) Milk contains the disaccharide lactose. **Figure 2** shows the biochemical pathway by which mammary gland cells produce lactose by a condensation reaction between glucose and galactose.

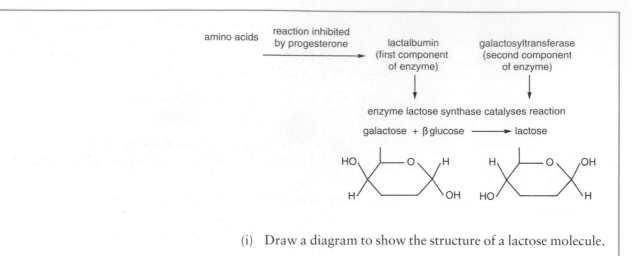

(i) Draw a diagram to show the structure of a lactose molecule.

(ii) Use the information in **Figure 2** to explain why there is an increase in lactose formation in the mammary glands immediately after birth.

*(2 marks)*

**Figure 3** summarises the mechanism involved in releasing milk from the mammary gland during sucking.

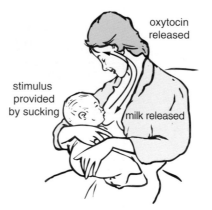

(c) This mechanism involves a reflex. For this reflex:

(i) where are the receptors

*(1 mark)*

(ii) what is the effector?

*(1 mark)*

(d) (i) Use information in **Figure 3** to suggest the precise action involved in the release of milk.

*(1 mark)*

(ii) When a child is suckling from its mother, the mother may experience contractions of her uterus. Explain why.

*(1 mark)*

Cancers occur in many different parts in the body but they occur more frequently in some organs than in others. If an error occurs during the replication of DNA a mutation results. The more mutations that occur, the more likely it will be that cancer will result.

(e)  Explain how mutation can result in cancer.

*(3 marks)*

(f)  (i)  Why are tissues such as those that form the mammary gland and the lining of the gut particularly susceptible to cancer?

*(2 marks)*

(ii)  There is a significantly lower chance of breast cancer occurring in women who have had children, women who reach puberty late and women who enter menopause early. Suggest why.

*(2 marks)*

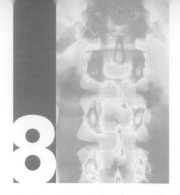

# Growth and Development to Maturity

Most of us have a body size and shape that is natural, i.e. we just grew that way. However, many people choose to change their body shape, sometimes by surgery.

Dieting to lose body fat is very common. The ex-Spice Girl, Geri Halliwell, made the tabloid headlines in 2001 when she went on a crash diet and the actor Sylvester Stallone regularly controls his diet to reduce the fat:muscle ratio of his body for his action films. Using a variety of exercise regimes to build muscle is also common. Some people build muscle because they prefer that appearance. Others build muscle because it helps with their job, e.g. as an actor or sportsperson.

Professional rugby union players are more effective in a highly physical game if they increase their upper body strength. They do this by a demanding exercise regime involving heavy weights. Those rugby players who are involved in catching a ball that is thrown into a line-out are more effective if they are tall. No amount of training can increase a person's height. Richard Metcalfe (Figure 8.1) made his debut for Scotland against England in the Calcutta Cup game of April 2000. At that time, Richard's body mass was 127 kg (20 stones) and was affected by a weight-training regime to increase his upper body strength. However, his standing height of 2.13 m (7 ft 1 in) is natural. This gives Richard a standing height advantage of about 10 cm over most of his competitors in the line-out. After reading this chapter, you will be able to define the terms body mass and standing height and to use your knowledge of the hormonal control of growth during childhood and adolescence to explain why Richard is so tall.

**Figure 8.1**
Although born in Leeds, Richard Metcalfe plays rugby union at international level for Scotland. His height of 2.13 m (7 ft 1 in) gives him a natural advantage in jumping for the ball in a line-out

# Definitions

Growth and development occur as dynamic, interrelated processes throughout the first twenty or so years of our lives. **Growth** involves a permanent increase in the amount of organic matter from which our bodies are made. This occurs in two ways: the production of new cells or an increase in the size of existing cells. **Growth hormone** (also called somatotropin) is released by the anterior lobe of the pituitary gland and stimulates both cell division and protein synthesis. **Development** involves predictable changes in the structure or functioning of our bodies as we mature and age. As well as physical changes, developmental changes include the emotional, mental and social changes that occur throughout our lives. Developmental changes are difficult to measure, but would include consideration of whether or not a child can walk unaided or how large a vocabulary a child has. Table 8.1 lists the main developmental stages through which we all pass.

| Stage | Description |
| --- | --- |
| Neonate | First two weeks after birth, by which time the neonate regains its birth mass. |
| Infancy | From second week to end of first year. Infant increases overall size and some organs, such as the brain, are well on the way to their adult size. |
| Childhood | From 1 to 12 years old. The brain achieves its adult size and the body proportions are similar to those of an adult. |
| Puberty | For girls, this occurs between the ages of 10 and 15 and for boys between the ages of 12 and 16. During this time, the sex organs and secondary sexual characteristics develop. |
| Adolescence | From puberty until about 3 years later. Behavioural and emotional changes, rather than biological changes, occur. |
| Adult | Bone formation and body growth are completed. In most systems, cell division occurs only to replace existing cells that have been lost. |
| Old age | Ageing results in general body deterioration and, finally, death. |

Table 8.1 The stages of human development

**Q 1 In which of the developmental stages shown in Table 8.1 does growth occur?**

## Measuring growth

Human growth is **diffuse**. This means that it occurs all over the body, rather than at a single growing point. We also grow in three dimensions, rather than in one dimension. An adult is taller than a child, but is also broader across the back, is larger from front to back and has larger organs throughout the body. We cannot measure all of these dimensions to get a picture of human growth. Instead, we use three key indicators, which are harmless, and easy, to measure.

- **Body mass** – During an ecological investigation, we could measure the amount of organic matter in samples of plants by drying them to a constant mass and then measuring their dry mass. We cannot do this with humans. We can measure wet body mass (commonly called weight) using a set of scales (Figure 8.2). This gives us an indication of the amount of substance in the body. Unfortunately, this method can be misleading since the measured mass includes food and water that were recently taken in by the person being measured. You can easily increase your body mass by eating a large meal and will lose body mass every night by losing water in your exhaled breath and in sweat as you sleep.

- **Standing height** – This involves a person standing shoeless and with their heels flat on the ground against a wall or measuring device. A horizontal bar is moved so that it touches the top of the person's head. The person's standing height is measured from the ground to the top of their head (Figure 8.3).

- **Supine length** – Standing height can be measured only in people who can stand. Most people do not learn to stand until they are about nine months old. Prior to this, an infant's length is measured instead. The infant is placed, on its back, on a table or floor. Its ankles are gently pulled to straighten the infant's legs and the length of its body, from the top of its head to the base of its heels, is measured.

**Q** **2** **If you are monitoring your body mass, why should you ensure that you weigh yourself at the same time of day?**

## Long-term studies of growth

Measurements of standing height and body mass are commonly used to monitor the growth of children. One of the first systematic attempts to record the growth of a child was carried out by a Frenchman, called Count Philibert de Montbeillard. He measured the height of his son every six months, from his birth in 1759 until his eighteenth birthday. Figure 8.4 shows a graph of the height of the Count's son during this time. This growth curve is typical. With the exception of two periods of rapid growth, the Count's son grew constantly throughout his life until his late teens, when the curve levels.

The Count de Montbeillard conducted a **longitudinal study**, meaning that he measured the same person over a number of years. This was easy to do, since he was measuring his own son. Longitudinal studies are

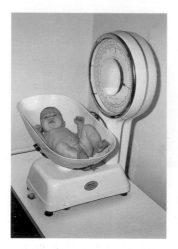

**Figure 8.2**
The body mass of this baby is being measured to monitor its growth

**Figure 8.3**
This child's growth is being monitored by measuring her standing height

**Figure 8.4**
The height of Count Philibert de Montbeillard's son from his birth in 1759 until his eighteenth birthday. This is an example of a longitudinal study of human growth

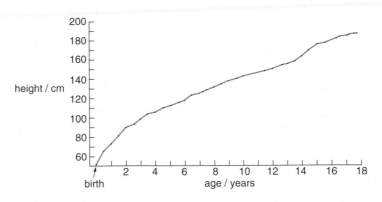

helpful because they give a very accurate picture of growth over the lifetime of the same people. In clinical studies, however, it is difficult for research workers to maintain contact with the same people throughout their lives. An alternative method is to use a **cross-sectional study**. Instead of measuring the standing height of the same people throughout their lives, in a cross-sectional study we would measure the standing heights of samples of people of different ages. For example, an alternative to the Count de Montbeillard's study would have been to measure the standing heights in a sample of one-year-old boys, in another sample of two-year-old boys and so on and then calculate the mean height of each sample. The mean values could then be plotted against age on a graph.

| Cross-sectional study | Longitudinal study |
|---|---|
| Measures samples of people of different ages and calculates mean for each age group. | Measures the same people at regular intervals throughout their lives. |
| Gives a generalised picture of growth, which does not relate to any one individual. As a result, individual peaks of growth tend to be smoothed out. | Gives a very accurate picture of individual growth. |
| Fails to compare like with like. For example, young people are generally taller today than 100 years ago. The mean height of 70–80 year-olds today relates to the mean height of 10–20 year-olds of 60 years ago and not to the mean height of 10–20 year-olds of today. | Since the same people are involved throughout, later measurements can be compared directly with measurements taken earlier in the study. |
| Easy to perform during clinical investigations since samples of people can be measured over a relatively short period of time. | Difficult to perform during clinical investigations since people move away from the study area or lose interest in being involved in the investigation. |

**Table 8.2** A comparison of cross-sectional and longitudinal studies

# Growth rates

A growth rate is calculated by subtracting the size measured at a particular time (T1) from the size measured some time later (T2) and dividing the result by the time interval. It can be expressed mathematically as:

$$\text{Growth rate} = \frac{(\text{size at T2} - \text{size at T1})}{(\text{T2} - \text{T1})}$$

The graph in Figure 8.4 shows **absolute growth,** in this case the actual cumulative height of the Count de Montbeillard's son. Like many curves of absolute growth, Figure 8.2 shows a fairly regular increase in size that begins to level out at about 16 years. The boy's absolute growth rate is shown in Figure 8.5. The **absolute growth rate** is a measure of absolute growth that has occurred within a specified time, in this case within each year. Figure 8.5 gives us much more information about the Count de Montbeillard's son. We can see that:

- the boy's growth rate was highest in the first year of his life

- his growth rate decreased rapidly during the first two years of his life

- he then grew at a fairly constant, low rate during most of his childhood

- he had a growth spurt in his mid teens

- he had almost stopped growing by the age of eighteen.

**Figure 8.5**
The growth rate of the Count de Montbeillard's son

**Figure 8.6**
(a) The mean height of samples of females and males of different ages
(b) The mean growth rate of females and males

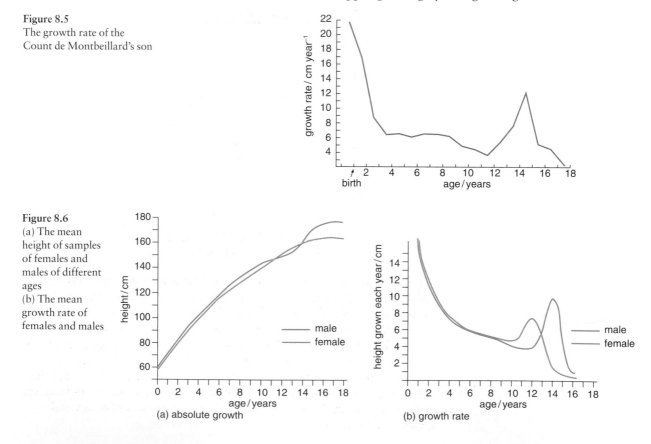

(a) absolute growth

(b) growth rate

Figure 8.6 shows graphs of absolute growth and of absolute growth rates for females and males. Notice in Figure 8.6(a) that females and males have similar heights until about 14 years, when males become taller. Figure 8.6(b) shows that this difference is the result of a growth spurt, which, although later in males than in females, is greater in males.

**Q** 3 **Do the data in Figure 8.6 represent a cross-sectional or longitudinal study? Explain your answer.**

Figure 8.7 represents growth in a different way. Instead of absolute growth rates, these curves show the size at a given time as a percentage of the final adult size. The curve for the whole body is similar to the curve in Figure 8.4. This is not surprising since height is an indicator of general body size. The curve for the brain and head is quite different to that of the rest of the body. This curve shows that the head and brain develop earliest, the brain being about 90% of its adult size by the age of 5 years. The reproductive system develops latest and remains below 20% of its adult size until puberty begins. The lymphoid tissue, which includes the appendix, spleen and thymus gland, reaches its maximum size before puberty when, under the influence of sex hormones, it reduces to its adult size.

**Figure 8.7**
Different parts of the body grow at different rates

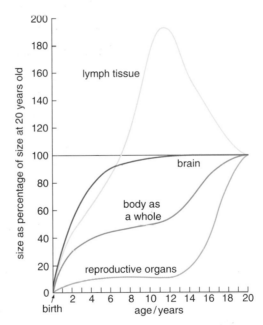

The different relative growth rates of body tissues and organs result in characteristic changes in shape during embryological and post-natal growth. Figure 8.8 shows the shape of a female body during growth. Two months after conception, the head of a fetus occupies about half its body length. At birth and for the following two years, the head occupies about one quarter of the body length whereas the head of an adult woman occupies about 13% of her height. We can conclude from Figure 8.7 that the change in proportions of the body shown in Figure 8.8 result from the early growth of the brain and head and the late growth of bones and muscles in the limbs.

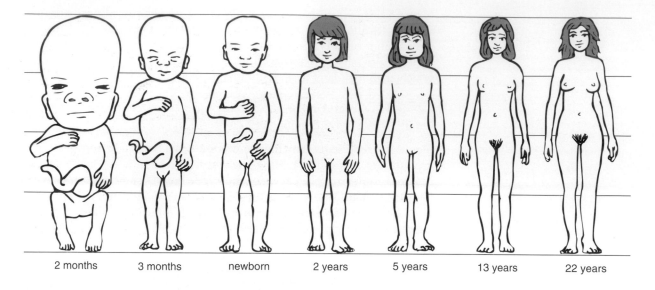

2 months    3 months    newborn    2 years    5 years    13 years    22 years

**Figure 8.8**
The proportions of the body change during development from prenatal conditions to adulthood

## Puberty

The growth spurts shown in Figures 8.4, 8.5 and 8.6 are associated with puberty. This is the time at which boys and girls become sexually mature and occurs earlier in girls than in boys. Figure 8.7 shows that, in addition to the general increase in size, the reproductive organs grow rapidly during puberty. There are, however, other changes that occur during puberty.

During childhood, there is little difference between the skeleton of a female and a male. At puberty, the hormone testosterone causes cartilage in the bones of a male's thorax and pectoral girdle to grow rapidly. As a result, a boy's shoulders expand in comparison with other areas of his trunk. In contrast, the hormone oestrogen causes the cartilage in a girl's pelvic girdle to grow rapidly. As a result, a girl's hips expand more than her trunk and shoulders.

Another important change that occurs during puberty is the relative amounts of protein and fat in the bodies of boys and girls. Testosterone in males causes a greater growth of muscle in boys than in girls. In contrast, oestrogen causes an accumulation of fat in the bodies of girls but not in boys. As a result, boys show a relative decrease in body fat as a percentage of total body mass whilst girls show a relative increase in body fat as a percentage of total body mass. As we will see later, the functioning of the menstrual cycle in females is related to the proportion of fat to muscle in a woman's body.

Perhaps the most dramatic changes that occur during puberty relate to the sex organs and sexual characteristics. The internal reproductive organs increase in size and change shape during puberty in both sexes. Associated with this, secondary sexual characteristics develop that are different in boys and girls. These are summarised in Table 8.3 and are caused by sex hormones, principally testosterone in boys and oestrogen and progesterone in girls.

| | Characteristic | Hormonal stimulation |
|---|---|---|
| **In boys** | Growth of testes | Interstitial cell stimulating hormone (ICSH), growth hormone and testosterone |
| | Growth of penis | Testosterone |
| | Growth of pubic hair | Testosterone |
| | Growth of axillary (underarm) and facial hair | Testosterone |
| | Body growth | Growth hormone, testosterone |
| | Growth of larynx, causing lowering of voice | Testosterone |
| | Sweat glands and sebaceous glands increase activity. This can result in acne, caused by blocked sebaceous glands | Testosterone |
| **In girls** | Growth of ovaries, oviducts, uterus and vagina | Follicle stimulating hormone (FSH), growth hormone and oestrogen |
| | Increase in diameter of pelvic girdle | Oestrogen |
| | Growth of breasts | Growth hormone, oestrogen and progesterone |
| | First menstrual flow (menarche) | Oestrogen and progesterone |
| | Growth of pubic hair | Oestrogen |
| | Growth of axillary (underarm) hair | Oestrogen |
| | Sweat glands and sebaceous glands increase activity. This can result in acne, caused by blocked sebaceous glands | Oestrogen |

Table 8.3
Physical changes that occur during puberty and the hormones that stimulate them

## Extension box 1

## Onset of puberty

The processes involved in puberty begin when a hormone, called **gonadotrophin releasing factor** (GnRF), is released from the hypothalamus. (You will learn more about the hypothalamus in Chapter 13.) This hormone stimulates the nearby pituitary gland to release hormones, called **gonadotrophins**. The main gonadotrophin is the same in males and females, but has a different name. In females, the gonadotrophin is called **follicle stimulating hormone** (FSH), because it stimulates egg-containing follicles that are found in the ovaries. In males, the gonadotrophin is called **interstitial cell stimulating hormone** (ICSH), because it stimulates cells that lie between the sperm-producing tubules in the testes. Gonadotrophins are carried in the blood stream and have a direct effect on the **gonads** (ovaries and testes), causing them to release **steroid sex hormones**. As you can see in Table 8.3, the main steroid sex hormones are oestrogen and progesterone in females and testosterone in males.

Although we know that the release of GnRF starts puberty, no one really knows what triggers the initial release of GnRF. Many mammals, such as the deer shown in Figure 8.9, have a seasonal reproductive cycle. In these mammals, GnRF release is triggered by day length, i.e. the proportion of dark and light hours during one day. It is likely that day length has an effect on the release of GnRF in humans too. However, GnRF is not released in pre-pubescent boys and girls, so some internal stimuli must also be involved in the onset of puberty.

One of these internal stimuli is related to diet. We know that the mean age of **menarche** (first period) is about four years earlier than it was in the nineteenth century. This is almost certainly related to improvements in diet, which enabled girls to grow more rapidly and reach the equivalent stage of maturity at a younger age. This influence of diet is related to the proportion of muscle to fat in females. Female athletes, who have a high muscle to fat ratio, tend to have a later menarche than girls with a lower muscle to fat ratio. Similarly, women who starve themselves during anorexia nervosa often find that their periods stop as they lose body fat.

Like many human processes, puberty is controlled by many factors.

**Figure 8.9**
Seasonal reproduction in these deer is controlled by day length

## Hormonal control of growth and development

Table 8.3 shows how puberty is stimulated by a number of hormones. Although many of these hormones appear only at puberty, others control growth throughout life. These hormones include growth hormone and thyroxine and are summarised in Table 8.4. In females, follicle stimulating hormone (FSH), luteinising hormone (LH), oestrogen and progesterone interact to control the regular events that occur during the menstrual cycle.

**Q**   4   **Explain why thyroxine will stimulate growth.**

| Name of hormone | Site of release | Principal actions |
| --- | --- | --- |
| Growth hormone (somatotropin) | Anterior lobe of the pituitary gland, situated just under the brain | Stimulates mitosis in body cells and stimulates protein production |
| Thyroxine | Thyroid gland, situated near the larynx (voice box) | Stimulates rate of metabolism |
| Pituitary gonadotrophins | Anterior lobe of pituitary gland | |
| ● Follicle stimulating hormone (FSH) | | In females, stimulates development of egg cells and secretion of oestrogen by ovaries. In males, stimulates sperm production in testes |
| ● Luteinising hormone (LH) | | In females, stimulates ovulation, formation of corpus luteum, thickening of uterus lining and milk production in breasts. In males, stimulates secretion of testosterone by testes |
| Oestrogen | Ovaries | In females, stimulates growth of secondary sexual characteristics (see Table 8.3) and thickening of uterus lining |
| Progesterone | Corpus luteum, formed in follicle after ovulation | Stimulates growth of breasts and increase in blood supply to uterus lining |
| Testosterone | Testes | In males, stimulates growth of secondary sexual characteristics (see Table 8.3) and sperm production |

**Table 8.4** Hormones involved in controlling growth and puberty

## Ageing (senescence)

**Extension box 2** — **The effect of ageing on collagen and elastin**

Collagen is one of the most common proteins, making up 30% of the total body protein. Together with elastin, it is found mainly in connective tissue, which holds other tissues together, and in the skin.

With ageing, the number of cells that produce collagen and elastin decreases. As a result, there are fewer fibres of collagen and elastin. The fibres that remain undergo structural changes, becoming larger and less resilient. Collagen fibres are made of three molecules of collagen. As Figure 8.10 shows, glucose molecules become attached to amino acids in the collagen molecules, eventually forming cross-linkages between the fibres. Elastin becomes calcified, making it less flexible.

**Figure 8.10**
During ageing, glucose molecules combine with free amino acids on collagen fibres, linking the fibres together. This contributes to wrinkles in the skin

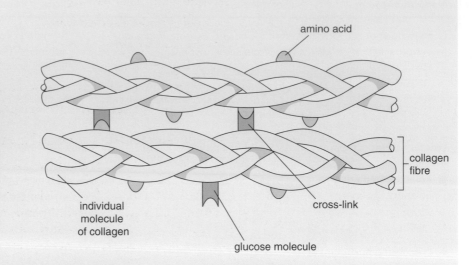

amino acid

collagen fibre

individual molecule of collagen

cross-link

glucose molecule

**Figure 8.11**
White hair and wrinkles in the skin are among the more visible signs of ageing

In the skin, the increase in size of collagen fibres reduces the space available for other molecules, such as lipids and water. As a result, skin becomes dryer as it ages. As elastin becomes less flexible, skin loses its elasticity, so that the skin loses its ability to smooth out. As a result, ageing skin sags and forms wrinkles. These are among the most obvious signs of ageing (Figure 8.11).

Ageing is a continuous process, beginning at birth and ending with death. Although we are all in one of several phases of ageing, the effects of ageing tend to become more obvious after about 40 years. Ageing is related to two main effects:

- **A reduction in the rate of cell division and, consequently, the number of cells** – Although all cells are capable of dividing during embryological development, some lose the ability to divide after birth. Neurones are one example of cells that cannot divide – we are born with a fixed number of neurones. If they are damaged or destroyed, they are not replaced. In cells that retain the ability to divide, such as epidermal cells, the rate of cell division slows with age.

- **A decline in the functional effectiveness of cells and of organ systems** – Deterioration in cells results in an ageing body becoming slower to respond to internal or external stimuli. In turn, this slows homeostatic mechanisms in the body, increasing the chance of dysfunction and death.

**Q  5  Explain why recovery from injury takes longer as we age.**

Although ageing is inevitable, the rate at which we age is highly variable. Our genes play a part, but the rate at which we age is also affected by choices we make about the way we live. Clinical studies have shown that six factors have a significant effect on physiological age. You will age more slowly if you:

- get regular and adequate sleep
- eat regular, well-balanced meals
- take regular exercise
- refrain from smoking
- either abstain from alcohol or drink alcohol only in moderation
- keep your body mass close to the desirable mass for your height.

## The effect of ageing on basal metabolic rate (BMR)

In Chapter 9, you will learn about the concept of basal metabolic rate, or BMR. In summary, the **basal metabolic rate** is the rate at which we use energy to carry out only the essential activities that keep us alive, such as transporting substances across cell membranes, breathing and pumping blood around our circulatory system. Other non-essential uses of energy, such as moving about, are not included in calculating the basal metabolic rate.

To remove the effect of overall body size, basal metabolic rate is measured as kJ kg$^{-1}$ hour$^{-1}$ (i.e. kJ of energy per kg of body mass per hour). Thus, the BMR depends on the number of metabolically active cells in a unit mass of the body. As we saw above, the number of cells decreases during ageing. Fewer active cells means a lower basal metabolic rate. In general, the basal metabolic rate decreases by about 5% every ten years above the age of 55.

**Q 6** What effect will an increase in the amount of stored fat have on the basal metabolic rate during ageing?

## The effect of ageing on cardiac output

Muscle cells become smaller if they are not exercised regularly. Cardiac muscle cells, especially those in the left ventricle, become smaller and may be lost during ageing. As a result, there is a progressive loss of cardiac muscle strength and a decrease in the stroke volume of the ventricles. As you learned in your AS course:

cardiac output = stroke volume × heart rate.

Consequently, a reduction in stroke volume results in the decrease in cardiac output shown in Figure 8.12.

**Figure 8.12**
A reduction in cardiac muscle strength causes the fall in cardiac output as we age

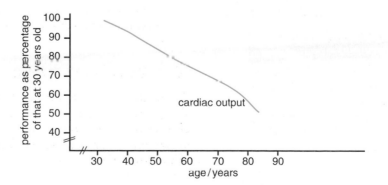

## The effect of ageing on the velocity of nerve conduction

The following changes occur in the nervous system as we age.

- Neurones (nerve cells) are lost. Since neurones cannot divide, this loss is permanent, so the number of cells in the brain and peripheral nervous system gets less.

- The rate at which impulses are conducted along neurones becomes slower. This results from the myelin sheath that surrounds axons becoming thinner, so that ions can leak out from nerve cells. You will learn more about the effect of myelin sheaths on impulse transmission in Chapter 11.

- Synaptic transmission becomes slower as presynaptic neurones produce less of their neurotransmitter. You will learn more about synapses in Chapter 11.

All these changes result in the decrease in velocity of nerve conduction that is shown in Figure 8.13.

**Figure 8.13**
Loss of nerve cells, thinning of myelin sheaths around individual nerve cells and loss of neurotransmitter at synapses contribute to a slowing of nerve conduction with age

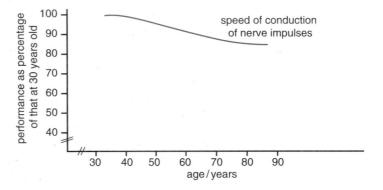

The effect of cell loss is dependent on where the cells were located. The brain normally loses about 25% of the cells that control muscular movement but hardly any of the cells that control speech. As a result, changes in muscle coordination are common as we age but changes in speech are not.

**Q** 7 Which of the effects of ageing, identified above, will be reduced by regular exercise? Explain why.

## The effect of ageing on female fertility

Although they suffer some loss of fertility, men generally remain fertile throughout their lives. From a biological point of view, it is, therefore, unremarkable that the actor David Jason recently became a father at the age of 61 or that the rock star Eric Clapton (Figure 8.14) became a father during the summer of 2001 at the age of 56. In contrast, women do not remain fertile throughout their lives. Sometime between the ages of 45 to 55, a woman's menstrual cycles become less and less regular and then stop altogether, so that she no longer ovulates. This is known as **menopause**.

**Figure 8.14**
Rock Star, Eric Clapton.

**Figure 8.15**
These are just a few of the anti-ageing products that are commercially available

Menopause occurs because women secrete less steroid sex hormones as they age, principally oestrogen. Figure 7.13 summarised the relationship between follicle stimulating hormone (FSH, released by the anterior pituitary gland) and oestrogen, released by the ovary. One effect of FSH is to stimulate the ovaries to produce oestrogen. However, at a critical blood concentration, oestrogen inhibits the further secretion of FSH. This is an example of negative feedback, which is so common in homeostasis (see Chapter 13). During menopause, a woman's ovaries become insensitive to FSH, so that they produce less and less oestrogen and, eventually, stop secreting oestrogen completely. As a result, a woman can no longer ovulate and so becomes infertile. Apart from this loss of fertility, a woman's vaginal walls become thinner and she might experience bouts of intense sweating and uncomfortable warmth ('hot flushes') and psychological problems during menopause. After menopause, a woman's risk of loss of bone tissue (osteoporosis) and of contracting heart disease also increases as a direct result of her low oestrogen levels. To counteract these symptoms, many postmenopausal women opt for **hormone replacement therapy** (HRT), in which they are given small doses of oestrogen and progesterone in tablets that they take orally, as implants beneath their skin or as skin patches.

**Q** 8 What will happen to the concentration of FSH in the blood of a woman after menopause?

## Extension box 3     Theories of ageing

Each year, people around the world spend the equivalent of billions of pounds on cosmetics, pharmaceuticals, herbal remedies and cosmetic surgery as they attempt to combat the effects of ageing (Figure 8.15). As highlighted earlier in this chapter, choices about our lifestyle might be more effective in reducing the rate at which we each age. However, one thing is certain: we will all age. With such interest in the ageing process, there are many theories to account for ageing.

The **gene mutation theory** suggests that gene mutations accumulate with age as we become less able to repair damaged DNA. As a result, abnormal cells are formed or dormant genes that were needed earlier in life become 'switched on' again. Such mutations could explain why cells begin to divide uncontrollably, giving rise to cancer.

The **chromosome mutation theory** suggests that we age because structural changes occur in our chromosomes. At the very ends of our chromosomes are lengths of DNA called **telomeres**. Each time a cell divides by mitosis, the telomeres get shorter. When its telomeres reach a critical length, a cell loses its ability to divide.

The **autoimmune theory** suggests that cell numbers decline because the immune system becomes less efficient, with two effects. Firstly, the immune system fails to recognise and destroy abnormal cells, so that they proliferate. These abnormal cells could become cancerous. Secondly, the immune system destroys healthy body cells, resulting in degenerative changes to tissues and organ systems.

The **hormonal theory** suggests that ageing is the by-product of a decrease in the number of hormone-secreting cells, resulting in a reduction in the secretion of hormones. In a recent clinical study in the United States of America, men over the age of 70 who were given small amounts of growth hormone showed a reversal in some of the symptoms of age, including an increase in the thickness of their skins and increases in the mass of their bones and muscles.

The **cellular garbage theory** suggests that ageing results from an accumulation of metabolic waste products inside cells. One group of metabolic waste products is called **lipofuscins**, which are oxidised lipids commonly found in ageing cells of the brain, heart and skin. These molecules cannot be broken down by normal enzyme action and accumulate in lysosome-like vesicles in the cytoplasm of cells. The accumulation of lipofuscins interferes with normal cell metabolism, either by directly interfering with normal metabolic reactions or by displacing other cell organelles.

The **free radicals theory** seeks to explain why substances in our cells become damaged with age. Free radicals are by-products of metabolism that contain unpaired electrons. As a result, they can oxidise and damage DNA, proteins and lipids. Lipofuscins, mentioned above, are lipids that have been oxidised by free radicals. Oxidised proteins, including enzymes, become common in ageing cells and the amount of oxidised DNA in mitochondria increases with age.

**Q** 9 **Explain why oxidised DNA in mitochondria could contribute to the effects of ageing.**

# Summary

- Growth involves a permanent increase in the amount of organic matter from which our bodies are made. Development involves predictable changes in the structure or functioning of our bodies as we mature and age.

- Growth is usually measured as supine length, in infants, standing height or body mass. Studies of growth can be longitudinal, measuring the same people throughout their lives, or cross-sectional, involving samples of people of different ages.

- Humans grow most rapidly in their first year, after which the rate of growth continues at a fairly constant, low rate. Puberty is associated with a brief growth spurt, which is greater and later in boys than in girls.

- Different parts of the human body grow at different rates. The brain and head reach their final adult size earliest whilst the reproductive system reaches its final adult size latest.

- During puberty, there is a general increase in size and the reproductive systems grow and mature; menarche occurs in females. Secondary sexual characteristics develop during puberty and include an increase in muscle in males but of fat in females, the growth of underarm and pubic hair and changes in behaviour.

- During childhood, growth is controlled by growth hormone and thyroxine. Puberty begins when pituitary gonadotrophins and steroid sex hormones (oestrogen and progesterone in females, testosterone in males) are secreted.

- Ageing occurs throughout life but is noticeable after about 40 years. There are many theories to account for ageing but it is associated with a decrease in the number of active cells in the body and a decline in physiological effectiveness of tissues and organs. Amongst the functions that are affected by ageing, the basal metabolic rate decreases, a decrease in stroke volume reduces cardiac output and the speed at which nerves conduct impulses gets less with age.

- Although men can father children throughout their lives, a woman loses her fertility during the menopause. This loss of fertility is caused by decreased secretion of pituitary gonadotrophins and of the ovarian hormones, oestrogen and progesterone.

# Assignment

## From child to adult

You will know from your own experience that different people grow and mature at different rates. The factors that govern these processes are complex. They are partly genetic and partly environmental. In this assignment, we will explore their influence on puberty in girls. The assignment is based on information collected from a number of studies, and will provide you with another opportunity to practise your data-handling skills. You will need to use material from various parts of the specification to answer the questions. Before you start, it would be a good idea to look up the following key topics, either in your notes or in your textbook:

● variation and its causes

● fertilisation.

We will start by looking at some of the hormonal changes that take place in a girl going through puberty. Look at the graph in Figure 8.16.

**Figure 8.16**
Changes in the concentrations of oestrogen and progesterone in the blood of a girl going through puberty

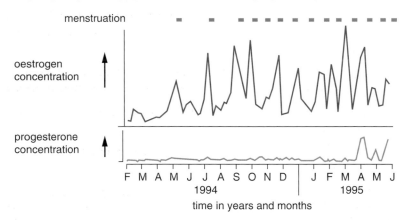

1 (a) The time of the first menstruation is known as menarche. Use the information in the graph to give the month and year of menarche in this girl.

*(1 mark)*

(b) Explain why it is easier to carry out studies concerning the onset of puberty in girls than in boys.

*(1 mark)*

2 The first ovulation usually takes place late in puberty. Suggest the biological advantage of this.

*(2 marks)*

3 (a) In what year and month did this girl first ovulate?

*(1 mark)*

(b) Explain the evidence from the graph that supports your answer to question 3(a).

*(2 marks)*

The age at which puberty begins varies from person to person. As with other aspects of development, the causes are partly genetic and partly environmental. One way in which we can investigate the influence of these two factors, is to compare the age of menarche in different pairs of girls. These pairs are identical twins, non-identical twins and sisters who are not twins. Table 8.5 shows the mean difference in age at menarche in girls from these groups.

Table 8.5

| Relationship | Number of pairs | Difference in age at menarche/months |
|---|---|---|
| Identical twins | 51 | 2.8 |
| Non-identical twins | 47 | 12.0 |
| Sisters who are not twins | 145 | 12.9 |

4   Explain why data like those in Table 8.5 should be analysed with a statistical test.

*(2 marks)*

5   Identical twins (Figure 8.17) are called monozygotic twins. Non-identical twins are called dizygotic twins. Explain why.

*(2 marks)*

6   Explain the evidence from the data in the table which supports the view that the age of menarche is partly determined by:

(a)  genetic factors

*(2 marks)*

(b)  environmental factors

*(2 marks)*

Figure 8.17
Identical twins are not only similar in appearance but they are also similar in many aspects of their growth and development

Studies have been carried out comparing the age of menarche in urban and rural areas throughout the world. The results of some of these studies are shown in Figure 8.18. This figure is adapted from work first published in 1976.

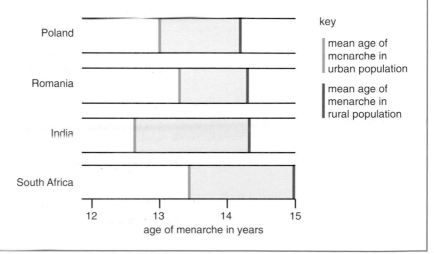

Figure 8.18

**179**

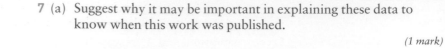

7 (a) Suggest why it may be important in explaining these data to know when this work was published.

*(1 mark)*

(b) What general conclusions can you draw from the data shown in Figure 8.18?

*(2 marks)*

8 Various ideas have been put forward to explain the difference between the age of menarche in urban and rural areas. They include:

● children in towns are subject to continuous illumination with artificial light

● rural children expend more energy in physical activity than urban children

● the total energy intake in urban diets is higher than that in rural diets.

Use information in Box 1 (page 170) to explain how each of these ideas could account for the difference in age at menarche.

*(6 marks)*

# Examination questions

1 The graph shows the rate of growth of the pelvis of a boy and of a girl as they reach maturity.

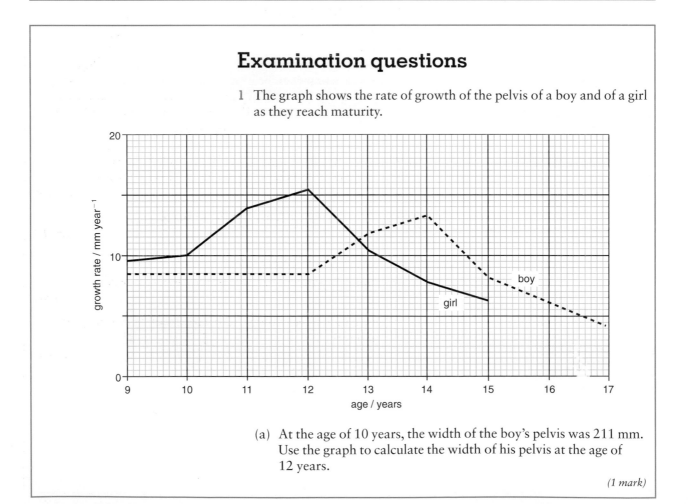

(a) At the age of 10 years, the width of the boy's pelvis was 211 mm. Use the graph to calculate the width of his pelvis at the age of 12 years.

*(1 mark)*

(b) Use the graph to describe the differences in growth rate of the pelvis which took place in these two individuals.

*(3 marks)*

(c) Explain how hormones account for the changes which take place in the width of the pelvis as boys and girls reach maturity.

*(3 marks)*

2 The figures from which the graph has been drawn were collected from a single group of boys. Curve **A** shows the rate of growth of one of the boys from the age of 10 years to the age of 18 years. Curve **B** represents the results of a cross-sectional survey of the whole group.

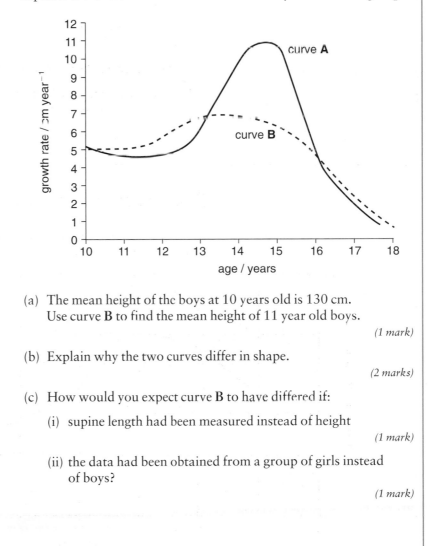

(a) The mean height of the boys at 10 years old is 130 cm.
Use curve **B** to find the mean height of 11 year old boys.

*(1 mark)*

(b) Explain why the two curves differ in shape.

*(2 marks)*

(c) How would you expect curve **B** to have differed if:

(i) supine length had been measured instead of height

*(1 mark)*

(ii) the data had been obtained from a group of girls instead of boys?

*(1 mark)*

# Digestion and Diet

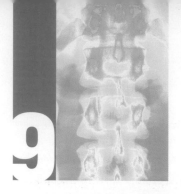

**Figure 9.1**
A tapeworm is a parasite which lives in the intestine of its host. It has a long, flattened body through which it absorbs digested food. This picture shows the hooks which the tapeworm uses to secure itself to the intestinal lining

Life must be very easy for a tapeworm (Figure 9.1). It is a parasite and spends its time lying in the intestine of its human host, surrounded by digested food. All it has to do is to absorb the products of digestion. It doesn't even need a gut of its own. Tapeworms, however, do have a number of adaptations which make them very efficient at absorbing the products of their host's digestion.

We will look at the body surface of a tapeworm. It is covered in cells like those shown in Figure 9.2. The surface of each of these cells has a large number of tiny, hair-like processes called microtriches. Each of these processes is about 2 mm in length and ends in a spine. Not only does the long, flattened shape of the tapeworm provide the animal with a large surface area but these processes increase it even more. The spines may help to hold the parasite in place in the intestine.

Although the tapeworm only secretes few digestive enzymes, it is now known that it is able to capture enzymes such as pancreatic amylase produced by its host. It absorbs these enzymes onto its body surface. It gains an advantage from being able to do this. Food molecules are broken down on the tapeworm's surface. This will help to make absorption more efficient as the distance the products have to travel to get into the body of the parasite is very small.

A tapeworm has another adaptation which allows it to compete very successfully with its host for the products of digestion. It secretes hydrogen ($H^+$) ions. This results in the pH of the intestine of an infected person being lower than that of an uninfected person. A lower pH makes the host intestine inefficient at absorbing molecules such as those of glucose, while glucose absorption by the parasite appears to be faster. The tapeworm's carrier molecules work better than those of the host at low pH values.

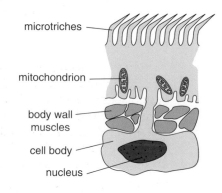

microtriches

mitochondrion

body wall muscles

cell body

nucleus

**Figure 9.2**
A cell from the body surface of a tapeworm. The microtriches provide a large surface area for absorption. Compare them with the microvilli shown in Figure 9.8. They have a similar function

The tapeworm and its human host are **heterotroph**s. They get their nutrients by breaking down large molecules such as carbohydrates and proteins into smaller ones. They do this, however, in different ways. In this chapter we shall concentrate mainly on the human digestive system.

There are four basic stages in the way in which humans process their food.

- **Ingestion** – Taking food in through the mouth.

- **Digestion** – Breaking down the large molecules which make up the food into smaller, soluble ones. This is the function of enzymes secreted in digestive juices produced by various glands.

- **Absorption** – Taking up the products of digestion. A variety of processes are involved here. They include diffusion and active transport.

- **Egestion** – The removal of undigested food, bacteria and dead cells from the lining of the gut as faeces.

**Q** 1 Excretion is the removal of waste products formed by biochemical reactions in the body. Why do we not refer to the removal of faeces as excretion?

We shall look at the processes of digestion and absorption and consider how the flow of digestive juices is controlled so that they are only secreted when there is undigested food present in the gut. We shall complete the chapter by looking at nutrition and considering how dietary requirements differ according to age, sex and occupation.

## Digestion

In order to function properly, the human body needs to be supplied with carbohydrates, triglycerides and proteins. These substances are all composed of large molecules, and large molecules, such as those of proteins and polysaccharides, are unable to cross plasma membranes. They cannot be taken up by cells lining the gut, so they cannot be taken directly into the body. Digestion involves breaking these large molecules down into their smaller components: proteins into amino acids; triglycerides into monoglycerides, glycerol and fatty acids; and starch into glucose. The chemical reaction involved is the same each time and involves breaking links between larger molecules by adding molecules of water. This reaction is hydrolysis and digestive enzymes are therefore all **hydrolases**.

The gut or **alimentary canal** consists of a tube running through the body from the mouth to the anus. Food is swallowed and travels down the **oesophagus** to the **stomach**. From the stomach it passes first through the **small intestine** then through the **large intestine**. Various glands add their secretions to the contents of the gut. These include the **salivary glands**, the **pancreas** and the **liver**. The arrangement of the organs which make up the digestive system is shown in Figure 9.3.

We will now look in more detail at the way in which different substances are digested.

**Figure 9.3**
The digestive system of a human.

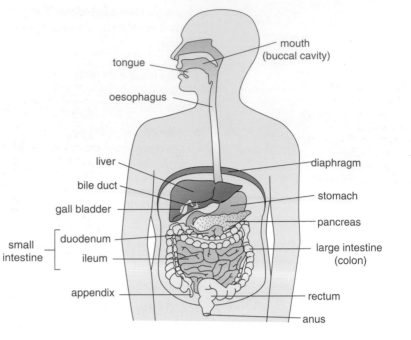

## Starch

Starch is a polysaccharide built up from a large number of glucose monomers. Digestion of starch involves two main steps: breaking down starch to maltose, and breaking down maltose to glucose.

The process starts in the mouth. Saliva contains **amylase**. Amylase is an enzyme which hydrolyses starch producing maltose as an end product. To be more accurate, salivary amylase is really a group of amylase enzymes, each of which hydrolyses starch by breaking the bonds at different places in the molecule.

**Q  2  Using your knowledge of enzymes, explain how different amylases hydrolyse starch by breaking bonds at different places in the molecule.**

Salivary amylase plays a relatively minor part in starch digestion. It works in slightly alkaline conditions. Food is rarely in the mouth for long and once it has been swallowed the pH changes rapidly as it encounters acid conditions in the stomach. In addition, much of the food we eat is hot. The temperature of a potato chip when it is eaten, for example, is likely to denature any amylase with which it comes into contact. Amylase is also secreted by the pancreas. This is a gland which opens through a duct into the small intestine. Most of the starch in the food we eat is broken down to maltose by pancreatic amylase.

The second stage in starch digestion – the hydrolysis of maltose to glucose – occurs in the small intestine. The enzyme that catalyses this reaction is **maltase**. Look at Figure 9.4 and you will see that the enzyme molecules are located in the plasma membrane of the epithelial cells which line the small intestine. Maltase breaks a molecule of maltose into two molecules of glucose. The glucose molecules are either absorbed into the cell or released into the lumen to be absorbed further along the small intestine.

**Figure 9.4**
Cells from the lining of the small intestine play an important part in the digestion of disaccharides such as maltose. The enzymes which catalyse this process are found in the cell surface or plasma membrane of these epithelial cells

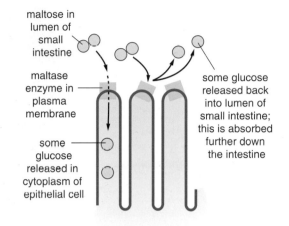

maltose in lumen of small intestine

maltase enzyme in plasma membrane

some glucose released in cytoplasm of epithelial cell

some glucose released back into lumen of small intestine; this is absorbed further down the intestine

## Lactose and lactose intolerance

Milk is often thought of as being the perfect food. It contains all the nutrients a young baby needs so why should it not be an ideal food for adults as well? Unfortunately for most of the world's adults, if they drink milk it makes them sick. The cause of this is the carbohydrate found in milk. It is a disaccharide called lactose. Babies and young children produce an enzyme in their small intestines called lactase. Lactase breaks down lactose into its two basic components, β-glucose and galactose, and these two monosaccharides can be absorbed into the body.

Many adults, however, no longer produce lactase. They are described as being lactose intolerant. If a lactose-intolerant person drinks milk, the lactose is not digested and cannot be absorbed from the small intestine. This has two effects. Lactose is soluble so the lactose remaining in the small intestine affects the water potential of the intestinal contents. It becomes lower (more negative). The cells lining the intestine have a higher water potential so instead of water being reabsorbed from the intestinal contents, it moves into them. The result is diarrhoea. In addition to this, the high concentration of lactose supports a large population of bacteria. They ferment the lactose and produce large amounts of gas, so a person who is lactose intolerant will suffer from diarrhoea and from the discomfort of a distended abdomen every time he or she drinks milk.

**Q** 3 **Yoghurt is produced by bacterial fermentation of milk. Suggest why people who are lactose intolerant are able to eat yoghurt.**

Lactose intolerance is found in a wide variety of mammals (see Figure 9.5). In humans, it is common in many parts of the world including much of Africa and tropical Asia. In view of this, it is quite surprising that the condition is relatively rare among people of West European origin.

**Figure 9.5**
Most adult cats are lactose intolerant so a saucer of milk is not a good idea. It is possible to buy lactose-reduced milk suitable for feeding to pets such as cats

## Extension box 1        Producing lactose-reduced milk

Because lactose intolerance is so common and widespread, there has been a lot of interest in developing industrial processes for reducing the amount of lactose in milk. There are commercial advantages in this because it would increase the potential market for dairy produce in areas like the Middle East. In addition, lactose-reduced milk could be very useful in making milk-based foods such as ice-cream. When stored for long periods in a freezer, ice-cream often develops a gritty texture. This is because lactose is not very soluble and crystallises in cold conditions. Lactose-reduced milk would make ice-cream with better keeping qualities.

**Figure 9.6**
An early reactor which used lactase to reduce the concentration of lactose in milk. The lactose was hydrolysed to the monosaccharides, β-glucose and galactose

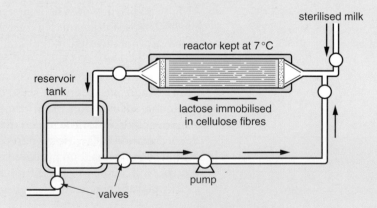

The ability to obtain large amounts of the enzyme lactase from a species of yeast, has made lactose-reduced milk a possibility. Figure 9.6 shows a reactor from a pilot plant designed and made in Italy. A batch of sterilised milk was fed into the reactor which contained lactase. The lactase was immobilised by packing it into cellulose fibres. Immobilising the enzyme like this had two important advantages over leaving it free in the solution. The product was not contaminated with lactase and it also helped to keep the enzyme stable – even after fifty consecutive batches of milk had been treated, there was only a 10% reduction in enzyme activity. The whole process was carried out at a temperature of 7 °C. This might seem rather cold for an enzyme-controlled reaction but it reduces the possibility of bacterial growth.

**Q** 4 **Explain why lactose-reduced milk tastes sweeter than fresh milk.**

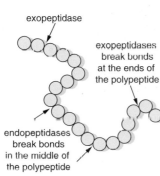

**Figure 9.7**
Polypeptides are long chains of amino acids linked by peptide bonds. Endopeptidases break bonds in the middle of the polypeptide while exopeptidases break bonds at the ends

## Proteins

Protein digestion is similar to the digestion of polysaccharides because it also involves hydrolysis. A protein molecule consists of one or more polypeptide chains. Protein-digesting enzymes or proteases break the peptide bonds joining amino acids together. The human intestine secretes a number of different **proteases** but they can be divided into two main sorts (see Figure 9.7).

- **Endopeptidases** break peptide bonds between specific amino acids in the middle of polypeptide chains. They break large polypeptides into smaller ones.

- **Exopeptidases** break peptide bonds between amino acids at the ends of polypeptide chains. They produce a mixture of amino acids and dipeptides.

**Q** 5 **As food passes along the gut, it is mixed first with endopeptidases and later with exopeptidases. Suggest why this arrangement is efficient at digesting proteins.**

Endopeptidases are secreted by glands, and glandular tissue contains a lot of protein … so, why do you not digest yourself? Why do these enzymes not break down the proteins of which your cells are made? Part of the answer to this question is that endopeptidases are secreted in an inactive form. Only when they get into the alimentary canal are they activated.

**Pepsin** is an endopeptidase secreted by the gastric glands in the stomach. It is secreted in an inactive form called **pepsinogen**. The stomach also secretes hydrochloric acid and this acid, together with pepsin already formed, converts pepsinogen to pepsin. This can be summarised in a simple equation:

$$\text{pepsinogen} \xrightarrow[\text{pepsin}]{\text{hydrochloric acid}} \text{pepsin}$$

Another endopeptidase is **trypsin**. Trypsin is secreted by the pancreas, again in an inactive form. This inactive form, **trypsinogen**, is converted to trypsin by another enzyme, **enterokinase**, which is secreted by the wall of the small intestine. Again we can summarise this in a simple equation:

$$\text{trypsinogen} \xrightarrow{\text{enterokinase}} \text{trypsin}$$

As well as producing trypsin, the pancreas produces a number of exopeptidases. They split groups of one or two amino acids off the ends of the smaller polypeptides which result from the action of pepsin and trypsin. The final stage in protein digestion occurs in the epithelial cells lining the small intestine. These cells have enzymes in their plasma membranes and in their cytoplasm which are able to convert dipeptides to amino acids.

## Lipids

Triglycerides are an important part of the diet. They are digested by lipase, an enzyme secreted by the pancreas, which breaks them down to produce glycerol, monoglycerides and fatty acids.

**Q 6** How many molecules are produced when lipase digests a molecule of triglyceride and produces:
(a) glycerol and fatty acids
(b) monoglyceride and fatty acids?

Vegetable oil is a triglyceride. If you shake a small quantity of vegetable oil vigorously with water, the oil breaks up into tiny droplets. These droplets are suspended in the water and form an **emulsion**. After a short time, however, they run back together to form a layer of oil. Lipase breaks down triglycerides much more effectively if they are dispersed to form an emulsion. With lots of very small droplets, there is a much larger surface area on which the enzyme can act. In the small intestine an emulsion of triglycerides is produced by the action of muscles and the presence of bile. Muscles in the wall of the stomach and the small intestine are continually mixing and squeezing the food. This produces an emulsion. The droplets do not run together again because of the presence of bile. Bile is made in the liver and stored in the gall bladder before being secreted into the small intestine. It contains a number of different substances including **bile salts**. It is these bile salts that make sure that the triglycerides stay as an emulsion.

**Table 9.1**
The main enzymes involved in digestion

| Part of gut | Secretion | Enzyme | Substrate | Products |
| --- | --- | --- | --- | --- |
| Mouth cavity | Saliva | Amylase | Starch | Maltose |
| Stomach | Gastric juice | Pepsin (endopeptidase) | Polypeptides | Small polypeptides |
| Pancreas | Pancreatic juice | Trypsin (endopeptidase) | Polypeptides | Small polypeptides |
| | | Exopeptidases | Polypeptides | Amino acids and dipeptides |
| | | Amylase | Starch | Maltose |
| | | Lipase | Triglycerides | Glycerol, monoglycerides and fatty acids |
| Wall of small intestine | | Maltase | Maltose | Glucose |
| | | Dipeptidase | Dipeptides | Amino acids |

## Absorbing digested food

Different parts of the alimentary canal have different functions. The role of the oesophagus is to squeeze food that has been chewed and mixed with saliva into the stomach. Once there, more mixing and churning take place and gastric juice starts the digestion of proteins. The small intestine is concerned with further digestion. It is also the region where most of the digested food is absorbed. Undigested food is stored in the large intestine before being egested from the body as faeces.

If you examine a section through any of these parts of the gut with a microscope, you can see that they all have a similar basic structure. They are hollow organs with a layer of epithelial cells surrounding the **lumen**, and their walls contain muscles and blood vessels. There are, however, slight differences in the structure of each region and these differences can be linked to function. Figure 9.8 shows the appearance of a cross section through the small intestine. Study this figure and you will see how the organ is adapted for its function of absorbing digested food.

**Figure 9.8**
The structure of the small intestine. The cross section (a) shows that the intestine is a tube with a thick wall surrounding a hollow lumen. When part of the wall is seen through a light microscope (b), the tissues that make up the organ are clearly visible. Higher magnification shows the epithelium in more detail (c). Photograph (d) is an electron micrograph of part of these epithelial cells

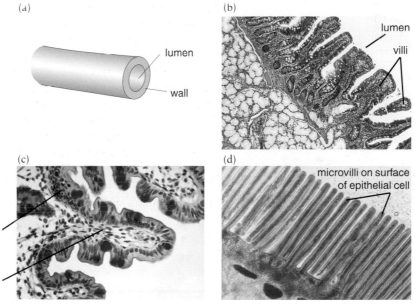

**Q** **7** List the features which result in the small intestine having a large surface area.

## Salts and water

By the time that the partly digested food is half way along the small intestine, it is rather like soup in consistency. It has been mixed with digestive juices from various glands and with sodium and chloride ions from the intestinal epithelium. These ions result in a water potential gradient. The water potential in the lumen of the first part of the small intestine is lower than that in the epithelial cells. Water therefore moves from these cells into the lumen by osmosis.

**Q** 8 **Which glands will have secreted digestive juices onto food by the time it is half way along the small intestine?**

Further along the small intestine, these ions are pumped back out again by the epithelial cells. This involves active transport. The result is that the water potential in the cells is now less than that in the lumen of the gut, so water moves back out by osmosis (Figure 9.9).

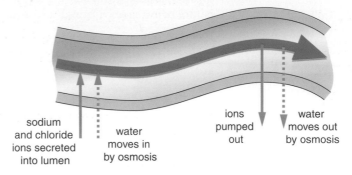

**Figure 9.9**
Sodium and chloride ions are secreted into the first part of the small intestine. This results in osmosis and water is added to the gut contents. The processes of digestion and absorption are much more efficient in the resulting soupy mixture. Further along the gut, the ions and the water are reabsorbed

## Glucose

The digestion of carbohydrates results in the formation of monosaccharides such as glucose. Glucose is absorbed from the gut by active transport. There are specific carrier molecules in the plasma membranes of the epithelial cells of the small intestine and these take the molecules into the cytoplasm of the cell. Uptake requires ATP and is linked to the transport of sodium ions. Glucose passes from the inside of the cell into the capillaries by facilitated diffusion (Figure 9.11).

## Oral rehydration therapy

For many people living in more developed countries, diarrhoea is only an inconvenience. It is unpleasant while it lasts, but it is usually over quickly. In many developing countries, however, diarrhoea can kill. Because their food supply is often inadequate, children living in parts of Africa, South America and Asia are particularly prone to infection. Infection further limits the nutrients they can absorb, so infected children become weaker and yet more prone to disease. The result is a vicious circle of malnutrition and diarrhoeal infection.

Infantile diarrhoea can be cured, however, as can various forms of diarrhoea affecting adults. What is more, the treatment is simple, cheap and effective. Before we look at this treatment – oral rehydration therapy (ORT) – we need to understand a little more about what causes diarrhoea. Many different types of microorganism may be responsible but frequently the large amounts of semi-liquid faeces result from toxins which they produce. Some of these toxins block the sodium channels and stop the reabsorption of sodium ions from the small intestine. As a consequence, the concentration of sodium ions in the small intestine increases. Water cannot be reabsorbed because the water potential gradient is now in the wrong direction. Instead, it is drawn out of the cells lining the small intestine and added to the contents of the gut. The result is diarrhoea.

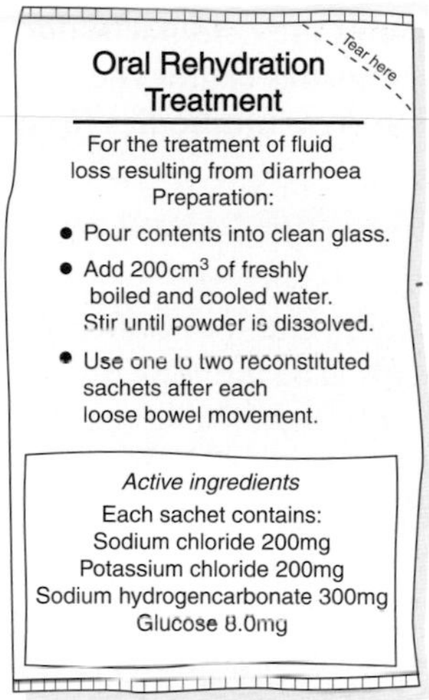

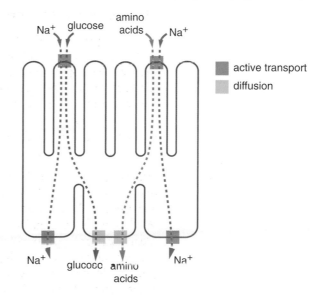

**Oral Rehydration Treatment**

For the treatment of fluid loss resulting from diarrhoea
Preparation:

● Pour contents into clean glass.

● Add 200cm³ of freshly boiled and cooled water. Stir until powder is dissolved.

● Use one to two reconstituted sachets after each loose bowel movement.

*Active ingredients*

Each sachet contains:
Sodium chloride 200mg
Potassium chloride 200mg
Sodium hydrogencarbonate 300mg
Glucose 8.0mg

**Figure 9.10**
Oral rehydration therapy is cheap and effective. The contents of this sachet may only cost a few pence but, correctly used, could save the life of a child

Although the toxins produced by these microorganisms prevent uptake of sodium ions, they have very little effect on the carrier proteins which transport glucose out of the small intestine. These proteins are known as cotransport proteins because they transport both sodium ions and glucose molecules. So, if we can get the cotransport proteins to work a little better, we should be able to ensure that adequate amounts of glucose and sodium ions pass into the intestinal cells and this ought to go a long way towards clearing up the attack of diarrhoea. This is the basis of ORT (Figure 9.10). Measured amounts of glucose and mineral salts are mixed with a set volume of clean water. Drinking the resulting solution stimulates sodium and glucose uptake by cotransport proteins. Water is absorbed from the small intestine and the attack of diarrhoea is brought under control.

**Q  9  Would adding more glucose to the ORT mixture make the treatment any more effective? Explain your answer.**

## Amino acids

The uptake of amino acids works in a slightly different way from that in which glucose is absorbed. This time, amino acids pass into the epithelial cells by facilitated diffusion. Different sorts of carrier molecule in the plasma membrane transport different types of amino acid. The carrier molecules only work, however, in the presence of sodium ions and each time an amino acid is transported into the cell, so is a sodium ion. A diffusion gradient is maintained across the plasma membrane by pumping the sodium ions actively out into the blood. Carriers also transport some dipeptides across the membrane. Enzymes break these dipeptides down to amino acids in the cytoplasm (Figure 9.11).

**Figure 9.11**
The ways in which glucose and amino acids are absorbed by the epithelial cells lining the small intestine. Note the importance of sodium ions in these processes

$Na^+$  glucose    amino acids    $Na^+$

active transport

diffusion

$Na^+$    glucose    amino acids    $Na^+$

Some proteins can be absorbed directly by the epithelial cells of the small intestine by pinocytosis. One example in which this occurs is in new-born mammals. They can absorb antibodies from their mother's milk in this way.

**Q 10** New-born babies do not produce trypsin and their stomachs do not secrete hydrochloric acid. Explain how these features enable the intestines of new-born babies to absorb proteins such as antibodies from their mother's milk.

## Lipids

We saw that triglycerides were digested by lipase enzymes to produce a mixture of glycerol, monoglycerides and fatty acids. The monoglycerides combine in the small intestine with bile salts to form tiny droplets called **micelles**. Each of these micelles is about 5 mm in diameter and also contains other molecules such as fatty acids and glycerol. The micelles transport their contents to the plasma membranes of the epithelial cells (Figure 9.12). They do not go into the cell, but their contents do. The glycerol, monoglycerides and fatty acids they contain dissolve readily in the phospholipid bilayer and enter the cytoplasm. Triglycerides are resynthesised and pass out through the sides and base of the cells into lymph capillaries called **lacteals**.

**Figure 9.12**
Bile salts have two important functions in digestion. They help triglycerides to form an emulsion (a) and they also help in transporting the products of triglyceride digestion to the epithelial cells lining the small intestine (b)

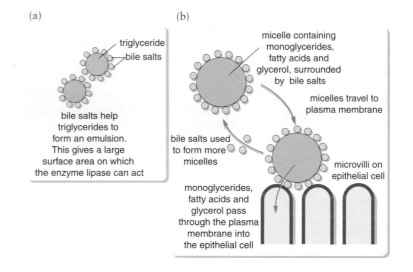

## Controlling digestion

Digestive juices are only secreted in large amounts when food is present in the gut. There are good reasons for this. Think, for example, about gastric juice. It contains hydrochloric acid and pepsin which is a protein-digesting enzyme. Prolonged contact with gastric juice would damage the tissues which form the stomach wall. In addition, secreting digestive enzymes when they are not required wastes the body's resources.

**Q 11** Name *two* substances which would be wasted if digestive juices were secreted when no food was present in the gut.

Figure 9.13 shows a baboon searching for tasty morsels in a pile of elephant dung. It does not know precisely when it is going to find an undigested seed or a maggot. Once it finds an item of food, the baboon puts it in its mouth, chews it and swallows it. It is necessary to have a

**Figure 9.13**
A feeding baboon. Food spends longer in the small intestine than it does in the mouth. It also takes longer to get to the intestine. These differences are reflected in the way that digestive juices are controlled

control system which ensures that the animal will have saliva in its mouth as soon as the food is put in. Food does not remain in the mouth for all that long, however, so the flow of saliva need only be short term.

The situation is rather different in the small intestine. This part of the gut is rarely empty and there is usually plenty of time before food put in the mouth arrives in the small intestine. A different sort of control system, one which takes these factors into account, is therefore needed for pancreatic juice.

A mammal has two main communication systems – the nervous system and the endocrine system. When you read Chapter 13 you will see that nervous communication relies on electrical impulses passing along nerve cells. Because of this, it is an ideal system for controlling processes which are short-lived and need to be switched on or off rapidly. The endocrine system is based on hormones travelling in the blood, from the glands or cells which secrete them, to the target organ. Hormones generally have long-lasting effects and take more time to bring about changes.

**Q 12** **Which of the nervous system or the endocrine system would you expect to control:**
**(a) a change in the diameter of the pupil when a light is shone at the eye**
**(b) the calcium concentration in the blood?**

There is a change of emphasis from nervous control to endocrine control as you move down the gut. Saliva is needed as soon as, or even just before, food enters the mouth. This is an ideal situation for nervous control. On the other hand, food takes a considerable time to reach the small intestine and is present there for some hours. Not surprisingly, secretion of bile and of pancreatic juice are mainly controlled by hormones.

There are three basic ways in which the nervous and endocrine systems control the secretion of digestive juices.

- **Nervous reflexes** – A reflex is a nerve pathway involving a small number of nerve cells – often only two or three. The response is automatic as it does not involve the front part of the brain. This means that a particular stimulus will always have the same effect. The other important feature of a reflex is that, as it only involves a small number of neurones, it is very rapid. You can find out more about reflexes on page 236.

- **Conditioned reflexes** – Has your mouth ever watered at the smell of your favourite meal cooking? This is an example of a conditioned reflex. The stimulus responsible for increasing the secretion of saliva is contact between food in your mouth and tastebuds on your tongue. But there are many sights, sounds and smells associated with meals. In time, one of these will make you salivate, even if you have no food in your mouth. We describe a conditioned reflex as when an unrelated stimulus – such as the sight or smell of food – produces the response which was originally associated with an appropriate stimulus.

- **Hormones** are secreted in response to the presence of food in a particular region of the gut. These hormones are carried in the blood to glands where they stimulate the secretion of digestive glands.

Table 9.2 summarises the way in which some digestive secretions are controlled in humans.

| Part of gut | Stimulus | Effect |
|---|---|---|
| Buccal cavity | Various stimuli associated with food such as the smell or sight of food, or sound of food being prepared | Trigger a conditioned reflex leading to the secretion of saliva before food arrives in the mouth |
| | Contact of substances in food with taste buds on tongue | Reflex leading to secretion of saliva when food is in the mouth |
| Stomach | Various stimuli associated with food | Trigger a conditioned reflex leading to the secretion of gastric juice before food arrives in the stomach |
| | Contact of substances in food with taste buds on tongue | Reflex leading to secretion of gastric juice as food arrives in the stomach |
| | Food in stomach | Secretion of the hormone gastrin. Gastrin travels in the blood to the glands in the stomach wall. These glands secrete gastric juice |
| Pancreas and gall bladder | Food in the small intestine | Secretion of the hormone cholecystokinin-pancreozymin (CCKPZ). This hormone stimulates the release of digestive enzymes by the pancreas and bile by the gall bladder |
| | | Secretion of the hormone secretin. Secretin stimulates the release of alkaline fluid by the pancreas |

**Table 9.2**
Summary of the ways in which some digestive secretions are controlled in humans

## What do we need to eat?

A balanced diet contains carbohydrate, lipid and protein. We need considerable amounts of these substances. In addition we need smaller amounts of minerals and vitamins. The roles of each of these substances in the body are summarised in Table 9.3.

You can see from the information in this table that food has two main functions. It provides the chemical potential energy which is the starting point for respiration and the substances it contains can be used in metabolism to produce or maintain the cells and tissues which make up the body. We will look at these two roles in a little more detail.

| Substance | Role in body |
| --- | --- |
| **Carbohydrates**<br>Starches and sugars | Provide up to 80% of our total chemical potential energy. Breast-fed infants obtain approximately 40% of their chemical potential energy from lactose, the sugar in milk. |
| Non-starch polysaccharides | Linked to control of appetite.<br>Diets low in non-starch polysaccharides (sometimes referred to as 'dietary fibre') have been related to a number of conditions including appendicitis, cancer of the colon, haemorrhoids and constipation. |
| **Lipids** | Together with carbohydrate, lipids are an important source of chemical potential energy. Phospholipids are an essential component of plasma membranes. Essential fatty acids are precursors of other important substances. |
| **Proteins** | Production of new tissues associated with growth and repair.<br>Enzymes and some hormones are also proteins and these play an essential role in metabolic processes. |
| **Minerals**<br>Iron | An important component of haemoglobin.<br>Many enzymes will only function if iron is present. |
| Calcium | Most of the calcium in the body is found in bones and teeth where it has a structural role.<br>A small amount is present in other tissues and in body fluids. Here it is important in transmitting signals. Its role in synapses is described on page 234. |
| **Vitamins**<br>Vitamin D | This vitamin is involved in the metabolism of calcium |

Table 9.3
The human diet

## Food and energy

Many of the foodstuffs we buy have a panel on the packaging which provides nutritional information (Figure 9.14). This panel shows the amount of energy in a standard sample. In semi-skimmed milk, for example this is 206 kJ per 100 ml. The figures are often given in both Joules and calories. As biologists we should use SI units. The SI unit for energy is the Joule (J). Because Joules are only small units, we normally use kiloJoules (kJ) or megaJoules (mJ) when referring to the amount of energy in food or required by an individual over a period of time.

1 megaJoule = 1000 kiloJoules

1 kiloJoule = 1000 Joules

**Figure 9.14**
The labels on many food products, such as this semi-skimmed milk, show nutritional information. This includes a reference to energy content

| NUTRITIONAL INFORMATION | | |
|---|---|---|
| TYPICAL VALUES | PER 200ml SERVING | PER 100ml |
| Energy | 411 kJ | 206 kJ |
|  | 98 kcal | 49 kcal |
| Protein | 6.8 g | 3.4 g |
| Carbohydrate | 10.0 g | 5.0 g |
| of which sugars | 10.0 g | 5.0 g |
| Fat | 3.4 g | 1.7 g |
| of which saturates | 2.0 g | 1.0 g |
| Fibre | nil | nil |
| Sodium | 0.1 g | trace |
| Calcium | 244 mg | 122mg |
| PER 200ml SERVING | 98 CALORIES | 3.4 g FAT |

**Q 13** Explain why the units of energy shown in the table in Figure 9.14 are given per 100 ml.

The simplest way to find the amount of chemical potential energy in a sample of food, such as a piece of pasta, is to burn a piece and use the energy released to heat a known volume of water. Figure 9.16 shows a simple way of doing this. The figure obtained by this method is not very reliable. There are a number of reasons for this. They include:

● a lot of the heat energy released on burning is not transferred to the water. It is lost to the surrounding air.

● the water in the boiling tube gets very hot. Some of this heat is also lost to the surrounding air.

● incomplete combustion of the pasta. A black, charred mass remains after burning and this still contains quite a lot of chemical potential energy.

**Figure 9.15**
Estimating the chemical potential energy in a piece of pasta

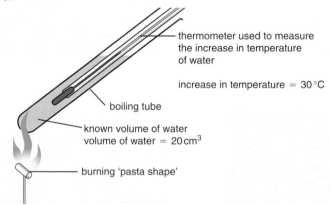

thermometer used to measure the increase in temperature of water

increase in temperature = 30 °C

boiling tube

known volume of water
volume of water = 20 cm$^3$

burning 'pasta shape'

**Q 14** Look at Figure 9.15. What would you need to do in order to express your results in:
(a) kJ
(b) kJ g$^{-1}$ of pasta

Although the principle remains the same, there are a number of more sophisticated ways of measuring energy content. One of these is shown in Figure 9.16. It produces a much more accurate result because:

- the water jacket completely surrounds the burning food sample. Very little heat energy is lost to the air

- the large volume of water in the jacket keeps the temperature rise small. Very little heat is lost from the water in the jacket to the surrounding air

- the food sample is burnt in oxygen. This ensures that combustion is complete.

**Figure 9.16**
This apparatus can also be used to estimate the chemical potential energy in a food sample. Although more complex than that shown in Figure 9.15, it relies on the same principles

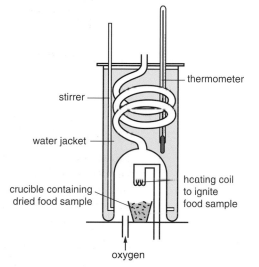

Foods rich in carbohydrates are often described as providing an 'instant' source of energy (Figure 9.17). This is because these carbohydrates are usually polysaccharides which are digested to produce glucose. Glucose is our main respiratory substrate. Carbohydrate-rich foods however tend to be bulky and contain a relatively large amount of water. A 100 g portion of boiled potatoes, for example, will contain 331 kJ of chemical potential energy. Compare this with a similar sample of a lipid-rich food such as butter. This will release 3100 kJ. A healthy diet requires the right balance between carbohydrates and lipids.

## Food and metabolism

You will know from your AS course that growth and maintenance of the body requires many different substances – nucleic acids carry genetic information; cells and tissues are built from proteins; reactions are controlled by enzymes and coenzymes; mineral ions are needed for the synthesis of many compounds and in carrying out functions such as the transmission of nerve impulses. All these substances come either directly or indirectly from the food we eat.

Proteins consist of polypeptides made of long chains of amino acids. There are 20 different amino acids commonly found in the proteins in our food and they fall into two groups. **Essential amino acids** are amino acids that we have to have in our diet because we cannot synthesise them from other substances. **Non-essential amino acids** on the other hand can be synthesised from essential amino acids. There is a list of essential and non-essential amino acids in Table 9.4.

**Figure 9.17**
Consuming food or drink rich in sugars is the fastest way to get a source of chemical potential energy into our bodies. They contain large amounts of glucose which can be absorbed without the need for digestion

| Essential amino acids | Non-essential amino acids |
| --- | --- |
| Isoleucine | Alanine |
| Leucine | Arginine |
| Lysine | Aspartic acid |
| Methionine | Asparagine |
| Phenylalanine | Cysteine |
| Threonine | Glutamic acid |
| Tryptophan | Glutamine |
| Valine | Glycine |
| | Proline |
| | Serine |
| | Tyrosine |
| | Tryptophan |
| | Histidine |

**Table 9.4**
Essential and non-essential amino acids

**Q** 15 **Is valine an essential amino acid because:**
**(a) it is essential in the diet**
**(b) it has an essential function in the body?**

Non-essential amino acids may be produced from essential amino acids in the liver. This involves a chemical reaction called **transamination** (Figure 9.18).

**Figure 9.18**
Transamination gets its name because it involves the transfer of an amino ($NH_2$) group from an essential amino acid to a molecule of another substance called a keto acid. This produces a non-essential amino acid and a molecule of a different sort of keto acid

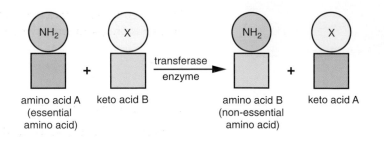

The term **vitamin** is used to describe one of a large group of substances which are not usually synthesised in the body so have to be obtained from the food we eat. They are only required in small amounts but play an important part in the metabolic reactions which take place in our cells. In this specification you only need to know about one of these substances, vitamin D.

## How much do we need to eat?

The food a human needs changes with age. It is not just the total amount that changes but the relative amounts of different nutrients. Table 9.5 shows how age affects the mean daily requirements for some nutrients in females.

| Age | Mean daily requirement for | | |
| --- | --- | --- | --- |
| | protein/g | iron/mg | calcium/mg |
| 0–3 months | 12.5 | 1.7 | 525 |
| 1–3 years | 14.5 | 6.9 | 350 |
| 7–10 years | 28.3 | 8.7 | 550 |
| 15–18 years | 55.5 | 14.8 | 800 |
| 19–50 years | 55.5 | 14.8 | 700 |
| over 50 years | 53.5 | 8.7 | 700 |

Table 9.5
Main daily requirements for some nutrients in females

Look at this table carefully. The amount of protein required can be linked with growth. Variation in the dietary requirement for calcium can be explained in terms of the growth of bones. A lot of calcium is required as the skeleton develops immediately after birth and there is another surge around puberty.

**Q 16** Iron is required for the formation of red blood cells. Suggest an explanation for the figures in Table 9.5 relating to the large amount of iron required by women in the 15–18 and 19–50 age groups.

In the rest of this chapter we will look more closely at the links between age, sex, occupation and our dietary requirements.

## The requirement for energy

### Basal metabolic rate

Asking how much food we need to eat is rather like asking how much petrol a car consumes. It depends on many different factors – the temperature of the surroundings, how active we are, age and sex. Just as in giving figures for petrol consumption, we need to standardise some of these variables if we want to make useful comparisons between individuals. This is where **basal metabolic rate** (**BMR**) comes in. BMR is the rate at which energy is released in a person who is at complete rest. It is a measure of the amount of energy required in a given time for essential activities such as the action of the heart and the contraction of muscles associated with breathing. BMR is measured some time after a meal so as to reduce as far as possible the effect of food in the stomach. It also has to be determined at a comfortable environmental temperature since processes such as shivering have a considerable effect on metabolic rate. BMR is particularly useful for measuring energy expenditure because:

- with the right laboratory facilities, it is quite easy to measure the BMR of an individual

- large numbers of measurements have been made. This has enabled us to produce a set of simple formulae. From these formulae BMR can be calculated knowing only a person's age, sex and body mass (Table 9.6)

- energy consumption at different levels of physical activity can be expressed in multiples of BMR.

You can use Table 9.6 to calculate your own BMR. Select the appropriate formula from the table. The figure you calculate will be your BMR in mJ per day.

| Sex | Age/years | Formula ($M$ = body mass in kg) | BMR for individual with body mass of 60 mJ |
|---|---|---|---|
| Male | 10–17 | $0.074M + 2.754$ | 7.194 |
| | 18–29 | $0.063M + 2.896$ | 6.676 |
| | 30–59 | $0.048M + 3.653$ | 6.533 |
| | 60–74 | $0.050M + 2.930$ | 5.930 |
| | 75+ | $0.035M + 3.434$ | 5.534 |
| Female | 10–17 | $0.056M + 2.898$ | 6.258 |
| | 18–29 | $0.062M + 2.036$ | 5.756 |
| | 30–59 | $0.034M + 3.538$ | 5.578 |
| | 60–74 | $0.039M + 2.875$ | 5.215 |
| | 75+ | $0.041M + 2.610$ | 5.070 |

**Table 9.6**
Calculating BMR in mJ per day from sex, age and body mass

A number of factors affect BMR. These include:

- **Body mass** – If we want to compare the basal metabolic rates of different individuals we must make some reference to size. The body of a large person contains many more cells than that of a smaller person. These cells are all respiring and releasing energy. One way of taking difference in size into account is to give BMR per kg of body mass.

- **Surface area** – One of the functions of basal metabolism is to supply the heat necessary to maintain body temperature above that of the surroundings. Since heat is lost from the body surface, it is not surprising that BMR is even more closely related to surface area than it is to body mass. So perhaps we should give BMR per $m^2$ of body surface? One of the main difficulties in doing this is that a person's surface area is very difficult to measure directly. It usually has to be found out from height and mass.

- **Sex** – Table 9.6 shows that there is a marked difference between the BMR of males and females of the same age. How can we explain this? The answer is concerned mainly with the fact that different tissues have different basal metabolic rates. In particular, muscle has a much greater BMR than fat. Men and women differ in body composition with men possessing a higher proportion of muscle and a lower proportion of fat than women (Figure 9.19). As a result, the BMR of a man is typically higher than that of a comparable woman.

**Figure 9.19**
One American study showed that, at the age of 25, the body of a man contained approximately 14% body fat while that of woman had around 25%. These figures increased with age. At 55, the corresponding figures were 26% for men and 38% for women

**Q 17  How would you expect pregnancy to affect BMR? Explain your answer.**

- **Age** – The basal metabolic rate of a new-born baby is low. It rises to a peak at one year of age and decreases steadily from then on. You can see this fall quite clearly by looking at the last column in Table 9.6. There are many factors which contribute to this pattern. Growth is a process which has very high energy demands. An actively growing child will therefore have a higher BMR, once difference in size has been taken into account, than an adult. The ratio of muscle to body fat also decreases with age and, as we saw in the previous paragraph, this also has a considerable effect on BMR. Finally, as we age, many of the biochemical reactions in the body slow down and become less efficient, further contributing to a fall in BMR. The changes in BMR due to differences in age and sex are summarised in the graph in Figure 9.20.

**Figure 9.20**
This graph summarises the effects of age and sex on BMR. Note the gradual decline with age and the fact that the BMR of females is lower than that of males across the entire age range

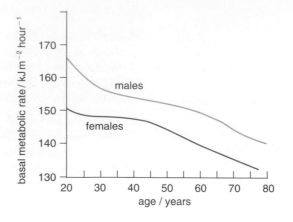

## Activity and metabolic rate

As we have seen, basal metabolic rate is a very useful concept. It has one big drawback, however, when we are considering our total energy requirement. Over much of the day we are physically active and BMR is only the metabolic rate at rest. One way of taking different levels of activity into account is to make use of the **physical activity ratio** (**PAR**). This is the ratio of the energy used in carrying out the activity concerned to the basal metabolic rate. Table 9.7 shows the PAR values of a number of different activities. If you study this table you will see that taking a shower, for example, would require 2.8 times as much energy as the basal metabolic rate.

| Activity | Approximate PAR value |
|---|---|
| Sitting down watching TV, writing or reading | 1.2 |
| Driving a car; carrying out a practical investigation in a laboratory | 1.6 |
| Standing and doing basic household chores such as cleaning or cooking | 2.1 |
| Dressing; taking a shower; walking at 3–4 km hour$^{-1}$ | 2.8 |
| Vigorous household chores such as cleaning windows or mopping the floor; walking at 4–6 km hour$^{-1}$; playing golf | 3.7 |
| Slow jogging, gentle cycling or walking at 6–7 km hour$^{-1}$ | 4.8 |
| Climbing stairs, playing football or tennis | 6.9 |

**Table 9.7**
Approximate PAR values for a variety of different activities

**Q** 18 A 17-year old girl has a BMR of 260 kJ hour$^{-1}$. How much energy would she require when playing a game of tennis?

## The requirement for protein

As with nutrients whose main function is to release the energy that a person requires, the amount of protein needed in the diet will also vary. Table 9.8 shows how the estimated average requirement for protein changes with age and sex. In order to get the figures in the table, the total daily requirement for protein has been divided by body mass. This enables us to make a better comparison as we do not have to take variations in body mass into account.

| Age/years | Estimated average protein requirement/g day$^{-1}$ kg$^{-1}$ | |
| --- | --- | --- |
| | Male | Female |
| 1–3 | 0.94 | 0.94 |
| 4–6 | 0.83 | 0.83 |
| 7–10 | 0.81 | 0.81 |
| 11–14 | 0.79 | 0.76 |
| 15–18 | 0.72 | 0.67 |
| 19–50 | 0.60 | 0.60 |
| 50+ | 0.60 | 0.60 |

Table 9.8
Protein requirements in males and females at different ages

We can use the figures in the table to illustrate some features of interest about the requirements for protein.

- The figures in Table 9.8 refer to egg or milk protein. For a vegetarian diet, we need to apply a correction factor. If the main sources of protein are cereal and vegetables, the figures in this table will need to be multiplied by 1.2. This is necessary to make sure that the required amounts of essential amino acids are eaten.

- When a person is growing, a lot of protein is required to produce new cells. Growth is very rapid in children and it is the youngest age group that requires most protein per kg of body mass.

- Growth is more or less complete after the age of 19. Nearly all the protein required in adults is used for maintenance and goes towards synthesising substances such as haemoglobin, enzymes and protein hormones. The amount of protein required for maintenance varies very little with age.

- A pregnant woman needs more protein in her diet. This is required for the growing fetus and for the growth of her own organs during pregnancy. A total of about 6 g of protein per day is required to meet these needs.

- Milk is rich in protein so extra is needed during lactation (Figure 9.21).

Figure 9.21
Extra protein is needed in order to produce milk. This is about 11 g per day during the first 6 months of lactation. It then falls to about 8 g per day after this as the protein content of milk decreases

**Q** 19 A 13-year old girl weighs 40 kg. Calculate the total amount of protein she requires per day if this came mainly from:
(a) animal sources
(b) cereal and vegetables.

## The requirements for iron and calcium

### Iron

Figure 9.22 shows the ways in which iron is taken into the body, is used or is lost. If you look at this diagram carefully, you should appreciate that some groups of people are particularly vulnerable to a deficiency of iron. These include:

- those with a high physiological requirement for iron. Rapid growth is linked with the need for iron in the diet. Young children and pregnant women therefore need a relatively high iron intake

- women of reproductive age. Menstruation results in a significant loss of blood and therefore iron (Extension box 2)

- children from cultural backgrounds in which late weaning and inappropriate infant food may create a problem

- elderly men and women, who often absorb iron poorly. People who drink too much tea may also suffer from this problem as the tannin in tea inhibits the absorption of iron from the gut.

**Figure 9.22**
The path of iron through the body. Low levels in the boxes shown in red and high levels in the boxes in blue will result in a deficiency of iron in the body

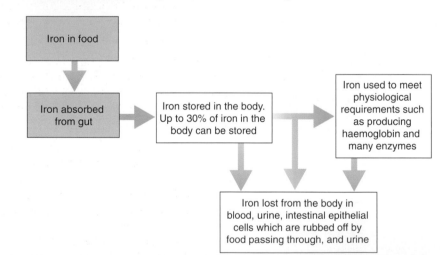

| Iron in food |
| Iron absorbed from gut |
| Iron stored in the body. Up to 30% of iron in the body can be stored |
| Iron used to meet physiological requirements such as producing haemoglobin and many enzymes |
| Iron lost from the body in blood, urine, intestinal epithelial cells which are rubbed off by food passing through, and urine |

**Extension box 2**    **Iron, menstruation and contraception**

Various factors affect the amount of blood a woman loses during menstruation. One of these is the presence or absence of an intra-uterine contraceptive (IUD). A study was carried out in which blood loss during menstruation was measured before and after IUDs had been fitted. The results are shown in Figure 9.23.

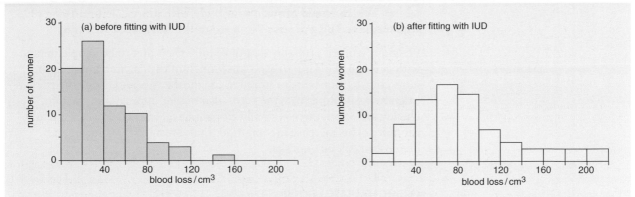

**Figure 9.23**
Histograms showing blood loss during menstruation in a group of 76 women (a) before and (b) after being fitted with IUDs

**Q 20** Describe the evidence from the histogram that the frequency of blood loss during menstruation does not show a normal distribution.

We will use figures from this study to calculate the amount of iron lost from the body as the result of menstruation.

- The mean blood loss during menstruation from this group of women, before IUDs were fitted, was 41 $cm^3$.

- We will assume that the concentration of haemoglobin in the blood is 13 g per 100 $cm^3$.

- A loss of 41 $cm^3$ of blood therefore represents a loss of $13 \times 41/100$ or 5.3 g of haemoglobin.

- 5.3 g of haemoglobin contains 18.4 mg of iron, so over a 28-day cycle, this would amount to an iron loss of 0.7 mg $day^{-1}$. Together with losses from other causes of about 0.8 mg $day^{-1}$, this is a total loss of 1.56 mg $day^{-1}$.

Now we will look at iron uptake into the body. In the UK, the average intake of iron per day for women between 18 and 49 is 12.1 mg. Of this, only about 15% is absorbed through the gut wall into the body, making the amount taken into the body about 1.8 mg $day^{-1}$. This compares with the loss of 1.5 mg $day^{-1}$ – a slight surplus.

Look at histogram (b). This shows the loss of blood during menstruation after IUDs had been fitted. Mean blood loss had increased from 41 $cm^3$ to 90 $cm^3$. Obviously iron loss and anaemia could present a real problem for women using IUDs, particularly if blood loss during menstruation was particularly heavy or the diet was low in iron.

**Q 21** IUDs appear to have many advantages as contraceptives for women living in developing countries. Use the information in this box to suggest a major disadvantage.

## Calcium

The amount of calcium taken into our bodies depends on the amount and the source. Watercress, for example contains 222 mg per 100 g of calcium while the same sized portion of semi-skimmed milk contains 122 mg. However, in one investigation it was found that 27% of the available calcium was absorbed from watercress but 46% was absorbed from semi-skimmed milk.

**Q** 22 **Which would provide your body with more calcium: 100 g of watercress or 100 g of semi-skimmed milk? Show your working.**

Pregnant women probably do not require their diet to be supplemented with calcium. This is partly because much of the calcium going to the fetus comes from calcium stored in the mother's bones. The efficiency of calcium absorption from the gut is also higher in pregnant women. During lactation, there is an additional factor. Lactating women tend to eat more. This increase in total food intake results in an automatic increase in calcium uptake.

## Special diets for special needs

### Weight loss diets

Obesity is a problem in many more-developed countries (Figure 9.24). Its immediate cause is very simple: energy intake in food is greater than energy expenditure in exercise. The difference between intake and expenditure can be very small. Suppose a person used a car every day for a short journey that could be walked in 15 minutes. This would amount to 250 kJ per day and could lead to a build-up of 10 kg of fat over 4 years.

Control of obesity means dealing with this energy imbalance and requires attention to both diet and to the amount of exercise undertaken. A weight loss diet needs to be constructed to suit the individual concerned but all these diets have features in common.

- Fat intake must be reduced because it: provides more energy than the same mass of carbohydrate; makes food tasty and therefore encourages eating; tends to be stored by the body rather than used as a respiratory substrate.

- A small amount of fat needs to be present, however, as it makes food palatable and provides essential fatty acids.

- Vitamin intake needs to be maintained. Plenty of fresh vegetables will enable the need for vitamins A and C to be met. Fresh vegetables also provide bulk which helps to decrease appetite and minimises the risk of constipation. Meat, fish, eggs and fruit should ensure enough of the other vitamins.

- Calcium needs can be met by drinking skimmed milk; the supply of iron, however, may need a supplement.

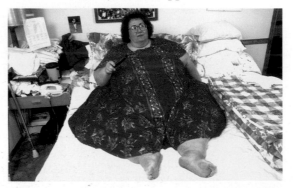

**Figure 9.24**
Obesity is defined as having too much body fat. It not only results in psychological problems associated with an unattractive appearance but leads to increased risk of diabetes, arthritis and heart disease

## Vegetarian diets

Those of us who eat meat and other animal products often do so because we like their taste. But they are not essential and it is perfectly possible to construct a balanced vegetarian diet. We have already looked at protein requirements and seen how a diet based on cereal grains and vegetables would have to contain more protein to meet the need for essential amino acids. Other problems with vegetarian diets include possible deficiencies of vitamin $B_{12}$ and vitamin D and, in some cases, a shortage of iron and calcium. Special attention needs to be given to getting adequate amounts of these substances.

## Glycogen loading and athletic performance

In the 1930s, experiments carried out in Sweden showed that if cyclists ate a diet high in carbohydrate three to four days before being tested, they could cycle for longer. Since then, other investigations have measured the effects of eating different amounts of carbohydrate on muscle glycogen. Figure 9.25 is a graph showing the results of one of these investigations carried out on a group of athletes. It shows that a high carbohydrate diet eaten in the days before running on a treadmill results in a much higher concentration of glycogen in the muscles. This increase in muscle glycogen does not allow a person to run any faster but it does allow him or her to keep going longer. On average, those on a low carbohydrate diet were exhausted after 59 minutes. Those on a high carbohydrate diet only became exhausted after 189 minutes. The more muscle glycogen, the greater the amount of glucose that can be made available for respiration and the longer the athlete can continue.

**Figure 9.25**
This graph shows the effects of low and high carbohydrate diets on muscle glycogen concentration and endurance time. See text for a detailed explanation

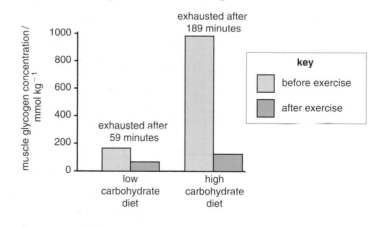

# Summary

- Digestion involves the hydrolysis of large insoluble molecules by the enzymes present in digestive juices and in the epithelial cells which line the small intestine.

- The small molecules resulting from digestion are absorbed through the wall of the small intestine. Absorption involves both diffusion and active transport. An understanding of the physiology of absorption enables us to control gastrointestinal infections by using oral rehydration therapy.

- The nervous and endocrine systems are involved in controlling the secretion of salivary, gastric and pancreatic juices.

- A balanced diet includes appropriate amounts of proteins, carbohydrates and lipids. Small amounts of vitamins such as vitamin D, and minerals such as calcium and iron are also required.

- Basal metabolic rate (BMR) is the rate at which energy is released in a person who is at complete rest. It is related to body mass and surface area and varies with age and sex.

- Physical activity increases metabolic rate and is reflected in the necessary food intake.

- Age and sex, pregnancy and lactation are factors which result in individuals requiring different amounts of energy-providing substances such as carbohydrates and lipid, proteins, mineral ions such as iron and calcium, and vitamins in the diet.

- The diet can be adapted to meet the demands of the individual concerned. Special diets for weight loss and vegetarians may produce potential nutrient deficiencies; athletes can eat high carbohydrate diets which improve athletic performance.

# Assignment

## Protein makes your wool grow

You have seen in this chapter how human nutritional requirements differ at different stages in their lives. In this assignment we will look at the protein requirements of sheep and how they differ. You will need to use material from various parts of the specification as well as from this chapter to answer the questions. Before you start, it would be a good idea to look up the following key topics either in your notes or in your textbook:

- proteins and protein structure

- genetic engineering.

It is important that farm animals are given the amount of protein they need. Too much would be wasteful and too little would slow down their growth. In order to be able to calculate the exact quantity required, we need a way of measuring the amount of protein in a food sample. One way of doing this is to calculate the crude protein (CP). It can be calculated from the equation:

$$CP = N \times \frac{1000}{160}$$

In this equation:

- N = the mass of nitrogen present in a 1 kg sample

- the ratio 1000:160 is based on the fact that 1000 g of protein contains 160 g of nitrogen.

1   The calculation of CP makes two assumptions. Explain why each of these assumptions could lead to an inaccurate result.

    (a)  All the nitrogen present in the food sample is present as protein.

*(1 mark)*

    (b)  All proteins contain 160 g of nitrogen per kilogram.

*(2 marks)*

Table 9.9 shows the different amounts of crude protein required in the diet of a sheep during pregnancy and lactation.

| Number of lambs in litter | Crude protein/g day$^{-1}$ | | | | | | |
|---|---|---|---|---|---|---|---|
| | Pregnancy/weeks before birth | | | | Lactation/weeks after birth | | |
| | 8 | 6 | 4 | 2 | 2 | 6 | 10 |
| 1 | 96 | 105 | 120 | 143 | 196 | 184 | 151 |
| 2 | 104 | 119 | 140 | 170 | 261 | 235 | 186 |

**Table 9.9**
Crude protein required by a sheep during pregnancy and lactation

2   (a)  Plot these data as a suitable graph.

*(4 marks)*

    (b)  Describe and explain the trend in the protein required in the diet during lactation.

*(2 marks)*

    (c)  The amount of protein required by the sheep changes during pregnancy. Suggest two functions for this extra protein other than providing protein for the growth of the lamb.

*(2 marks)*

3   Some of the protein in the diet of the sheep may be described as rumen degradable protein (RDP). Use your knowledge of digestion in ruminants to suggest the meaning of this term.

*(2 marks)*

Sheep's wool consists mainly of the protein keratin. The total mass of wool produced by a sheep in a year contains 3 kg of keratin.

4   (a)  Assuming that wool grows evenly, a sheep will need to synthesise approximately 8 g of wool keratin per day. Explain how this figure is calculated.

*(1 mark)*

    (b)  For each gram of protein in its food, a sheep can digest and absorb 0.54 g. How much protein would it need to eat in order to synthesise 8 g of wool keratin? Show your working.

*(1 mark)*

(c) Keratin contains $120\,\text{g}\,\text{kg}^{-1}$ of sulphur-containing amino acids. Plant protein contains $30\,\text{g}\,\text{kg}^{-1}$ of these amino acids. Use these figures to make a better estimate of the amount of plant protein a sheep would need to eat in a day in order to synthesise its wool keratin.

*(1 mark)*

Australian scientists have recently produced genetically-engineered clover containing a protein high in sulphur-containing amino acids. It is thought that feeding sheep on this new strain of clover will improve the yield of wool. A piece of DNA was prepared with the sequences shown in Figure 9.26. This piece of DNA was inserted into a clover plant.

5 Describe how each of the following enzymes may be used in preparing this piece of DNA:

(a) ligase

*(1 mark)*

(b) restriction endonuclease.

*(1 mark)*

6 Briefly explain why it is necessary to:

(a) ensure that the protein is made in leaf cells

*(1 mark)*

(b) prevent the protein from being digested in the rumen of the sheep.

*(2 marks)*

gene coding for protein high in sulphur-containing amino acids

gene ensuring that protein is made in leaf cells

gene which prevents protein from being digested in rumen

**Figure 9.26**

# Examination questions

1 (a) The diagrams show fatty acid molecules in saturated and unsaturated triglycerides.

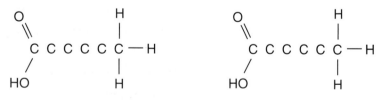

Complete the diagrams to show the difference between a saturated fatty acid and an unsaturated fatty acid.

*(2 marks)*

(b) In the human gut, a triglyceride may be converted to glycerol and fatty acids by hydrolysis.

(i) Explain what is meant by *hydrolysis*.

*(1 mark)*

(ii) Describe the part played by bile in the hydrolysis of triglycerides.

*(2 marks)*

2  Read the following passage.

Milk provides the essentials for life – carbohydrates, proteins, lipids and minerals. The raw materials, derived from the mother's blood, are taken up by the cells lining the alveoli of the mammary gland. These cells make or transport all the constituents of milk and secrete them into the
5  lumen of the alveoli from which ducts conduct milk to the nipple.

Milk contains a range of unique proteins, the caseins, which contain a high proportion of essential amino acids. Caseins precipitate in the acid pH of the stomach and form a protein clot which is then digested in the stomach. Enzymes hydrolyse the peptide bonds,
10  releasing amino acids for absorption into the blood. Casein also has an important role in transporting the calcium which is needed by the growing baby. The molecules of casein associate together with mineral ions to form structures known as micelles.

Milk also contains other molecules such as α-lactalbumin. Together
15  with a common enzyme called galactosyl transferase, α-lactalbumin makes a new enzyme complex, lactose synthetase. It is this enzyme which enables the mammary gland to make lactose from its constituent monosaccharides. This disaccharide is the main carbohydrate in milk and it is also extremely important in drawing water into the secretion.

20  Milk is not simply a source of food, it also has a high concentration of antibodies. The gut epithelium of a new-born mammal is able to take up these large proteins intact. Some classes of immunoglobulins are transported into milk from the blood. This transport across the mammary epithelium is often selective. Immunoglobulin A is made
25  locally in the mammary gland by B lymphocytes of the immune system. Remarkably, these B lymphocytes migrate to the mammary gland from special areas of the gut where they have already been exposed to antigens. As a result, the baby receives protection against the pathogens to which the mother has already been exposed.

*Source*: adapted from *Biological Sciences Review*, September 1997

*Using information in the passage, and your own knowledge, answer the following questions.*

(a)  Explain the meaning of the following terms as used in the passage:

   (i)  alveoli (line 3);

*(1 mark)*

   (ii)  enzyme complex (line 16).

*(1 mark)*

(b)  (i)  Describe a biochemical test by which you could demonstrate the presence of protein in milk.

*(2 marks)*

   (ii)  Describe the functions of **three** of the proteins which are found in milk.

*(3 marks)*

(c) Suggest the biological importance of the following statements.

   (i) Casein contains a high proportion of essential amino acids (lines 6–7).

*(2 marks)*

   (ii) Caseins form a protein clot in the stomach (lines 8–9).

*(2 marks)*

(d) Milk contains a very high proportion of water. Explain how water is drawn into the secretion produced by the mammary gland (line 19).

*(2 marks)*

(e) In some parts of the world diarrhoea is very common among infants. Suggest two reasons why the incidence of diarrhoea is lower in breast-fed babies than in those which are bottle-fed.

*(3 marks)*

3 Haemoglobin is an oxygen-carrying pigment found in red blood cells. A molecule of haemoglobin consists of an iron-containing haem group and a protein called globin. The diagram shows how haemoglobin is produced in the human body.

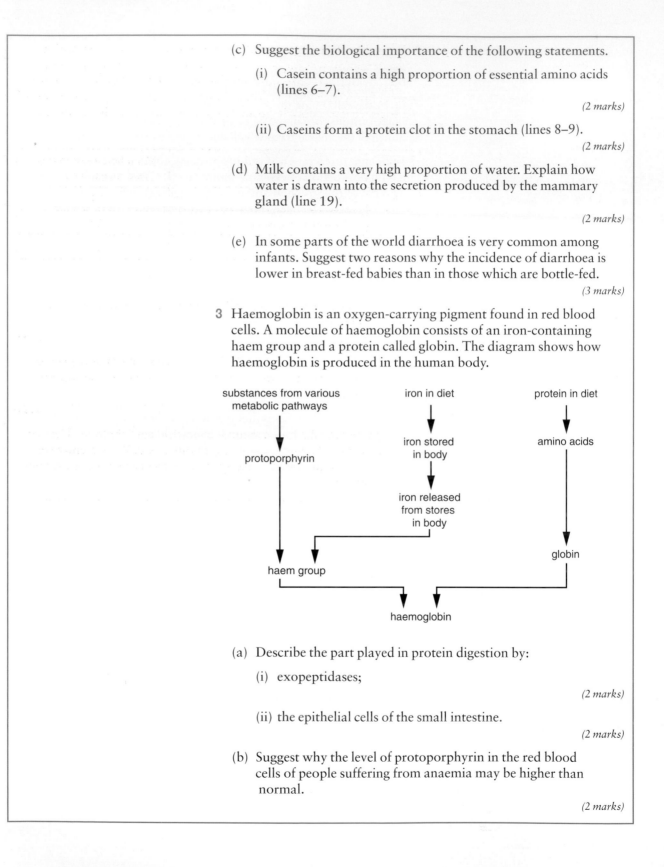

(a) Describe the part played in protein digestion by:

   (i) exopeptidases;

*(2 marks)*

   (ii) the epithelial cells of the small intestine.

*(2 marks)*

(b) Suggest why the level of protoporphyrin in the red blood cells of people suffering from anaemia may be higher than normal.

*(2 marks)*

In a study of three groups of women, the mean blood loss during menstruation was measured. Each of the women in group **A** was fitted with a standard intra-uterine device (IUD). Those in group **B** were fitted with an IUD which continually released small amounts of progesterone. Those in group **C** acted as a control. The results are shown in the table.

| Group | Treatment | Mean blood loss during menstruation/$cm^3$ |
|-------|-----------|---------------------------------------------|
| A | Standard IUD | 91 |
| B | IUD which releases progesterone | 19 |
| C | Control group | 41 |

(c) Suggest an explanation for each of the following observations:

  (i) anaemia is often associated with the use of intra-uterine devices;

  *(1 mark)*

  (ii) the mean blood loss during menstruation of the women in group B was lower than that of the women in the control group.

  *(3 marks)*

(d) There are 12.5 g of haemoglobin in 100 $cm^3$ of blood. The iron content of haemoglobin is approximately 0.3%. Calculate the mean amount of iron lost by the women in group **C**. Show your working.

# Transport of Respiratory Gases

Whichever way you look at it, 9000 metres or 29 000 feet, Everest is a very impressive mountain. Although it has been climbed many times, getting to the top is still a challenge. Not only is it bitterly cold, but the atmospheric pressure is far lower than at sea level. This brings real problems for the climber. The air at these high altitudes contains the same proportion of oxygen – around 21% – as it does at sea level but, because of the low pressure, there are far fewer molecules present. This means that there will be less oxygen going into the lungs, less oxygen being transported by the red cells in the blood and less oxygen being delivered to the muscles and other tissues in the body.

**Figure 10.1**
The environmental conditions encountered by a climber on Everest are very different to those found at sea level. Given time, the human body is able to adapt to these conditions. Many of these adaptations involve the blood system and the transport of oxygen to the tissues

One of the most remarkable features of the human body is that it can adapt to different environmental conditions, even when they are as demanding as those faced by mountaineers. These changes, however, do not occur immediately (Table 10.1) and mountaineers must spend time at high altitude to acclimatise. This helps the body to adjust to the demanding environmental conditions.

| Change | Time scale |
|---|---|
| Increase in resting heart rate<br>Increase in resting breathing rate | These are the first responses to living at high altitude. They start to occur after a few minutes. |
| Increase in concentration of blood plasma | A medium-term change which takes a week or so. |
| Increase in red blood cell production<br>Increase in number of blood capillaries | Long-term changes beginning to become noticeable after 1–2 weeks at high altitude. |

**Table 10.1**
The changes which take place as a mountaineer acclimatises

A red blood cell differs from other cells in the human body. It does not have a nucleus and there are no mitochondria in its cytoplasm. Instead, it is packed with around 280 million molecules of **haemoglobin**. In this chapter we shall look at the remarkable properties of haemoglobin which allow a mere five litres of blood to transport all the oxygen the human body needs. We shall also look at the other side of the respiratory equation. How is the carbon dioxide produced during respiration removed from the body? You may find it useful to read Chapter 7 in the AS book again before you start work on this chapter.

## Haemoglobin and the transport of oxygen

A molecule of haemoglobin from the blood of an adult human is made up of four subunits (Figure 10.2). Each of these subunits consists of two parts – **haem** and **globin**. The haem part consists of a ring of atoms linked to iron ($Fe^{2+}$). The globin part is a polypeptide chain.

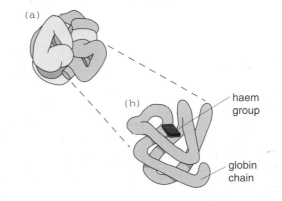

**Figure 10.2**
(a) There are four subunits in a molecule of haemoglobin from an adult human. The two blue ones are the α subunits. The two yellow ones are the β subunits. The subunits differ from each other in having slight differences in the sequences of amino acids which form the globin chains; (b) shows one of these subunits enlarged. The haem and globin parts of the subunit have been labelled

**Q 1 What is the evidence from Figure 10.2 that haemoglobin has a quaternary structure?**

There are many different sorts of haemoglobin. One way in which they differ from each other is in the sequences of amino acids which form the globin chains. This difference affects the oxygen-carrying properties of the molecule (Figure 10.3).

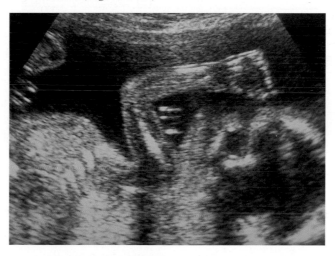

**Figure 10.3**
An ultrasound scan showing a human fetus inside the uterus of its mother. It has a slightly different type of haemoglobin from that found in an adult. This means that its haemoglobin is able to pick up oxygen from the mother's blood in the placenta

It is haemoglobin that results in blood being so efficient at transporting oxygen. In the lungs haemoglobin combines with oxygen to form **oxyhaemoglobin**. Oxyhaemoglobin is transported in the blood to the respiring tissues where it dissociates. It gives up oxygen which diffuses into the cells of the body while the haemoglobin is transported back to the lungs. To understand this process in more detail, we need to look at the **dissociation curve** shown in Figure 10.4. This is a graph showing the percentage saturation of a sample of haemoglobin plotted against the **partial pressure** of oxygen. The partial pressure of oxygen, usually abbreviated to $pO_2$, is simply a measure of the amount of oxygen present.

**Figure 10.4**
A dissociation curve for human haemoglobin. This graph shows how the percentage saturation of haemoglobin with oxygen varies with the amount of oxygen present

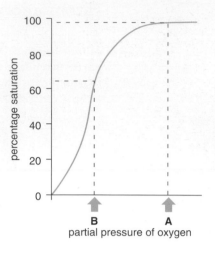

We will look at this graph in a little more detail. Find the point marked **A** on the horizontal axis. This point represents the partial pressure of oxygen in the lungs. You can see from the curve that haemoglobin is almost totally saturated at this high partial pressure. Now find the point marked **B**. This corresponds to a muscle. The muscle is respiring so it is using up oxygen. There will obviously be a lower partial pressure of oxygen in an actively respiring muscle than there is in the lungs. The curve shows that the percentage saturation of haemoglobin is lower at this lower partial pressure of oxygen, so a lot of the oxygen that was being carried by the haemoglobin will be given up.

So, to sum up – the haemoglobin in the blood is almost saturated with oxygen in the lungs. This oxygen is then transported by the blood to the tissues of the body where it is unloaded.

**Q** 2 (a) **What will happen to the partial pressure of oxygen in a muscle if its rate of respiration increases?**
(b) **What effect will this change in partial pressure have on the amount of oxygen supplied to the muscle?**

Another feature that you should note from Figure 10.4 is that the curve is S-shaped. The explanation for this lies in the fact that each haemoglobin molecule can carry up to four molecules of oxygen. The first oxygen molecule only binds with some difficulty but, as it does, it brings about a change in the molecular shape of the haemoglobin allowing it to bind more easily with the remaining oxygen molecules. This means that the last oxygen atom binds to the haemoglobin several hundred times faster than the first one.

The graph you have just been looking at shows what happens when there is very little carbon dioxide present, in other words, when there is a low partial pressure of carbon dioxide. Look at Figure 10.5. This shows you how a change in the partial pressure of carbon dioxide will affect the dissociation curve for haemoglobin. Increasing the partial pressure of carbon dioxide causes the curve to move over to the right. We refer to this effect as the **Bohr shift**.

**Figure 10.5**
The effect of changing the partial pressure of carbon dioxide on the dissociation curve for human haemoglobin

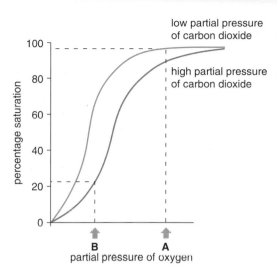

It is important to appreciate that Figure 10.4 represented a simplification. In the body not only does the partial pressure of oxygen vary but so does the partial pressure of carbon dioxide. Bearing this in mind we will now look at the graph in Figure 10.5 in more detail.

**Q**   **3**   **Would you expect to find the highest partial pressure of carbon dioxide and the lowest partial pressure of oxygen in the lungs or in a respiring muscle? Explain your answer.**

Find the point marked **A** on the horizontal axis again. As we saw above this represents the partial pressure of oxygen in the lungs. But now we have the extra complication of carbon dioxide to think about. In the lungs we will have a low partial pressure of carbon dioxide because it is being removed from the body. We are dealing with the upper curve and you can see from this that haemoglobin is almost totally saturated.

Now we will look at what is happening in the respiring muscle at the point marked **B**. The muscle is using up oxygen so there will obviously be a low partial pressure of oxygen. It is also producing carbon dioxide as a result of respiration so there will be a high partial pressure of carbon dioxide. We need to look at the lower curve which represents the situation with a high partial pressure of carbon dioxide. The effect of increasing the partial pressure of carbon dioxide therefore results in the haemoglobin giving up even more of the oxygen it is carrying to the tissues.

If you think about your own activity pattern, you will realise how important it is to match the amount of oxygen the blood supplies to the needs of the body. Going from standing still to walking quickly will result in a faster rate of respiration and the need for more oxygen to be supplied to the tissues. Two very simple mechanisms help haemoglobin to meet this need by releasing more of the oxygen it is carrying. As the rate of respiration increases more oxygen is used up and more carbon dioxide is produced. The resulting fall in the partial pressure of oxygen and rise in the partial pressure of carbon dioxide both lead to the release of more oxygen.

### Extension box 1

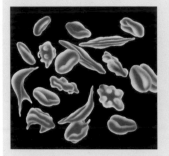

**Figure 10.6**
The blood cells in this photograph were taken from a patient with sickle-cell anaemia. Note their characteristic sickle shape

## Mutations and malaria

As you have just seen, a molecule of adult human haemoglobin is made up of four subunits – two α subunits and 2 β subunits. The sequence of amino acids in the globin chain of the α subunit is different from that in the β subunit. The two sequences are controlled by different genes. The gene controlling production of the α subunit globin is on chromosome 16 while that controlling the β subunit globin is on chromosome 11.

Mutations sometimes occur in these genes and we now know about over 400 different ones which affect the globin chains of human haemoglobin. Most of these involve substitutions of single DNA bases which affect single amino acids. Although this might not seem important, some of these mutations have very serious consequences. We will look at one of them – the one that gives rise to a condition known as **sickle cell anaemia**.

Sickle-cell haemoglobin differs from normal adult haemoglobin by just one amino acid. The sixth amino acid in the β subunit globin is valine, not glutamic acid. This results in sickle-cell haemoglobin being less soluble in conditions where there is a low concentration of oxygen, such as in the blood which is being carried back to the heart in the veins. The abnormal haemoglobin molecules stick together to form long fibres which alter the shape of the red blood cells, making them sickle shaped (Figure 10.6).

The body rapidly destroys these abnormal cells so an affected person develops severe anaemia. In addition, sickle cells clump together and may block smaller blood vessels. This reduces the supply of oxygen to the organs of the body. Not surprisingly a person who is homozygous and has two alleles for the condition is likely to die in early childhood unless he or she receives the necessary medical treatment. You might expect as a result of natural selection that the allele which causes sickle-cell anaemia would be very rare. This is not the case, however. In Africa, the condition is much more common than might be expected. This is particularly true of areas in which malaria is found (Figure 10.7). How can we explain this?

**Figure 10.7**
Maps of Africa showing the distribution of (a) the sickle-cell allele and (b) malaria. Note how high frequencies of the allele occur where malaria is common

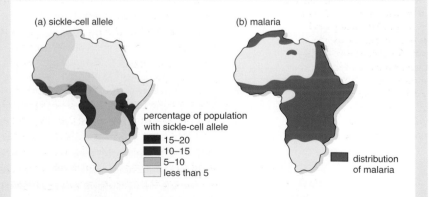

**Figure 10.8**
Malaria is a disease which affects the blood. This photograph shows red blood cells infected with malarial parasites. Eventually the cells will burst. A new generation of parasites will be released and infect more red blood cells

In order to understand the link between the sickle-cell allele and malaria we need to look at what happens in people who are heterozygous and only have a single copy of the defective allele. They make some sickle-cell haemoglobin and some normal haemoglobin. As a result they show few symptoms of sickle-cell anaemia although in some conditions, such as at high altitude and while carrying out vigorous exercise, some of the red blood cells become sickle shaped.

**Q** 4 **Explain why vigorous exercise causes some of the red blood cells in a heterozygote to become sickle shaped.**

People who are heterozygous have an advantage if they get malaria. The malarial parasite (Figure 10.8) causes the red blood cell it has infected to become sickle shaped. As a result, this blood cell is destroyed by the body and the parasites it contains are unable to complete their life cycles.

## Different sorts of haemoglobin

**Figure 10.9**
This is the end of the Earth! It is a photograph of the dramatic scenery at the southern tip of South America. In an area such as this the amount of oxygen available to different animals is very variable. The partial pressure of oxygen is particularly low on the top of the mountains in the background and in the burrows of the many small mammals that escape the extreme conditions by living underground. In addition, the whales and seals that live in the sea also have adaptations that enable them to remain underwater for up to an hour without breathing

Different animals live in different places with different environments. These environments are characterised by features such as temperature and humidity. A desert, for example, usually has a much greater temperature range and a lower relative humidity than a tropical forest. Environments also vary in the amount of oxygen that is available to the organisms living in them. The partial pressure of oxygen will be low, for example in deep underground burrows, on high mountains, and in lakes full of decomposing vegetation. Animals that live in these places have haemoglobin which is adapted to these conditions. This enables them to survive in conditions where the partial pressure of oxygen is low (Figure 10.9). Another organism that lives in an environment which has a low partial pressure of oxygen is the human fetus.

### Living where the amount of oxygen is low – the human fetus

In Chapter 7 you read about the placenta. You will have seen that a fetus gets its oxygen by diffusion from the blood of its mother. This blood has been pumped from the capillaries in the mother's lungs. Before it reaches the placenta it gives up some of the oxygen it was carrying to organs such as her uterus. Because of this, blood arriving at the placenta has a lower partial pressure of oxygen than that in the pulmonary veins.

The fetus, therefore, lives in a place where oxygen is in limited supply. It has haemoglobin of a different type to that found in adults. Figure 10.10 shows the dissociation curves for fetal and adult haemoglobin. Notice how the position of the fetal haemoglobin curve differs from that for adult haemoglobin. This enables fetal haemoglobin to become saturated with oxygen at the low partial pressures that exist in the uterus.

**Figure 10.10**
Dissociation curves for adult haemoglobin and fetal haemoglobin. Fetal haemoglobin is saturated with oxygen at lower partial pressures than adult haemoglobin

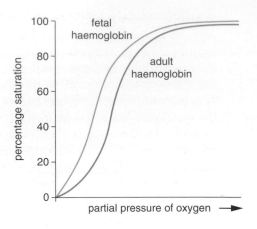

## Getting rid of carbon dioxide

As a result of respiration, cells produce carbon dioxide and this has to be taken from the tissues to the gas exchange surface where it is removed from the body. Carbon dioxide dissolves in water. It also reacts chemically with water to form carbonic acid. Carbonic acid is a weak acid but, under the conditions found in the body, it splits into hydrogen ($H^+$) ions and hydrogencarbonate ($HCO_3^-$) ions. This process can be summarised with an equation.

$$CO_2 + H_2O \rightleftharpoons H_2CO_3 \rightleftharpoons H^+ + HCO_3^-$$

| carbon dioxide | water | carbonic acid | hydrogen ion | hydrogencarbonate ion |

We will look at what happens in respiring tissues. Carbon dioxide is being produced. This diffuses from the cells into the blood plasma. A small amount, about 10%, reacts with the water in the plasma and produces hydrogen ions and hydrogencarbonate ions. Most of the rest diffuses straight into the red blood cells where it also reacts with water. Red blood cells, however, contain an enzyme called **carbonic anhydrase**. This enzyme catalyses the reaction so it proceeds much faster in red blood cells than in the plasma (Figure 10.11).

**Figure 10.11**
Carbon dioxide produced as a result of respiration reacts with water in the plasma and in the cytoplasm of red blood cells. As a result, hydrogen ions and hydrogencarbonate ions are formed

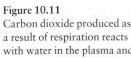

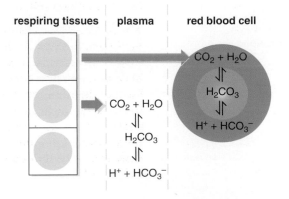

## Transporting hydrogencarbonate ions

Because the reaction between carbon dioxide and water is faster in the red blood cells than in the plasma, there is a higher concentration of hydrogencarbonate ions in a red blood cell. As a result these ions diffuse out through the cell surface membrane into the blood plasma. Ions, however, are charged particles and movement of all these negatively charged hydrogencarbonate ions into the blood plasma will upset the equilibrium. There is a mechanism which corrects this. The plasma contains a higher concentration of chloride ($Cl^-$) ions than the red blood cells. The hydrogencarbonate ions diffusing out are matched by chloride ions diffusing in. This mechanism is sometimes called the **chloride shift**.

Although most of the carbon dioxide is transported in the blood as hydrogencarbonate ions, a small amount – again, about 10% – reacts directly with haemoglobin to form **carbamino-haemoglobin**. The amount of carbon dioxide that haemoglobin can carry in this way depends on the amount of oxygen that it is carrying. The smaller the amount of oxygen, the more the carbon dioxide it can carry.

**Q** 7 Would you expect more carbon dioxide to be carried as carbamino-haemoglobin in an artery or in a vein? Explain your answer.

## Taking up the hydrogen ions

The more carbon dioxide that diffuses into the blood, the more hydrogen ions are formed. If the concentration of hydrogen ions were allowed to increase, the pH of the blood would fall. This would have severe effects on the biochemical reactions which take place in the body by affecting proteins such as enzymes, and carrier molecules found in plasma membranes.

You may remember that when you investigated the various factors which influenced the rate of enzyme-controlled reactions, you made use of **buffer solutions**. A buffer solution is one that keeps the pH of a solution constant by binding to hydrogen ions when their concentration is high and releasing them when it is low. The blood contains a number of substances which act as buffers and take up the hydrogen ions. In the plasma this function is carried out by phosphates and by plasma proteins. In the red blood cell the most important buffer is haemoglobin.

The faster a tissue respires, the more carbon dioxide it produces. As we have seen, this carbon dioxide combines with water to produce hydrogencarbonate ions and hydrogen ions. So, clearly, there are more hydrogen ions produced at greater rates of respiration. In red blood cells, most of these hydrogen ions are taken up by the haemoglobin. This, in turn, causes the haemoglobin to release the oxygen that it is carrying. So, the faster the rate of respiration, the more carbon dioxide produced and the greater the amount of oxygen supplied to the tissues. The way in which haemoglobin transports oxygen and carbon dioxide is summarised in Figure 10.12.

Figure 10.12
A summary of the way in which oxygen and carbon dioxide are transported in the blood

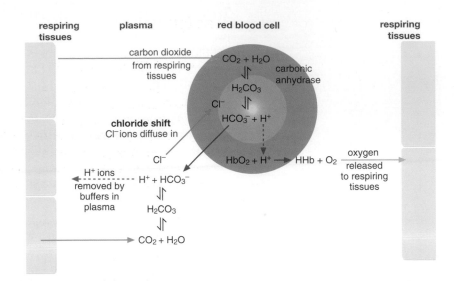

# Summary

- In mammals and many other animals, oxygen is transported from the gas exchange surface to the respiring tissues mainly by haemoglobin. When the concentration of oxygen is high, such as in the lungs, haemoglobin combines chemically with oxygen to form oxyhaemoglobin. When the oxygen concentration is low, such as in actively respiring tissues, the oxyhaemoglobin dissociates and oxygen is released.

- An oxygen dissociation curve is a graph showing the relationship between the percentage saturation of haemoglobin and the partial pressure of oxygen. The presence of carbon dioxide alters the position of the curve on the graph. This is known as the Bohr shift.

- Different organisms have different sorts of haemoglobin which have different oxygen-carrying properties. A human fetus has a special sort of haemoglobin which is saturated with oxygen at relatively low partial pressures.

- Most of the carbon dioxide produced in respiration is transported to the gas exchange surface as hydrogencarbonate ions.

- Haemoglobin is an important buffer. It absorbs the hydrogen ions produced by the reaction of carbon dioxide and water. As a result the pH of the blood remains more or less constant.

# Assignment

## Size and oxygen supply

You have read in this chapter that different animals have different sorts of haemoglobin with different oxygen-carrying properties. In this assignment we will look at other sorts of mammal and investigate how their size is related to the properties of their haemoglobin. You will need to use material from various parts of the specification to answer the questions. Before you start, it would be a good idea to look up the following key topics either in your notes or in your textbook:

- surface area to volume ratio

- metabolism and metabolic rate

- mitochondria and respiration

Although the mammals are a group of animals that are similar in many ways – their skin is covered in hair, they possess sweat glands, and they feed their young on milk – they differ considerably in size. The largest mammal, the blue whale, weighs about a hundred tonnes. This is about two hundred million times heavier than the smallest shrew! Look at the graph in Figure 10.13. It shows the oxygen dissociation curves for some different mammals.

**Figure 10.13**
The oxygen dissociation curves for a number of different mammals. They range in body mass from an elephant weighing about 5 tonnes to a mouse with a body mass of approximately 15 g

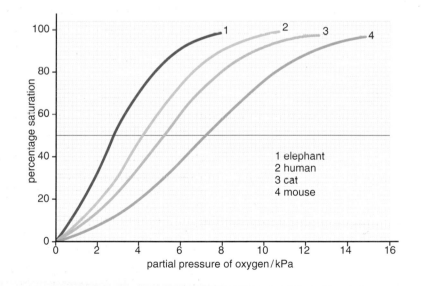

When we compare oxygen dissociation curves it is often useful to look at what is called their **unloading pressure**. This is the partial pressure of oxygen at which the haemoglobin is 50% saturated with oxygen.

1   Use Figure 10.13 to find the unloading pressure of the haemoglobin from each of the four mammals shown in the graph. Write your answers in a suitable table.

*(2 marks)*

The unloading pressure gives us a measure of how readily a sample of haemoglobin gives up the oxygen that it is carrying. The higher the unloading pressure, the more readily the haemoglobin gives up the oxygen it is carrying to the tissues.

2 Describe the relationship between the size of a mammal and the ease with which oxygen is unloaded from its haemoglobin to its tissues.

*(1 mark)*

Now, how can we explain this? How do small mammals benefit from haemoglobin that gives up its oxygen very readily? In order to answer this question, we need to look at another aspect of mammalian physiology. Look at the data in Table 10.2. This shows some of the results of an investigation carried out over a hundred years ago into heat lost by dogs with different body sizes.

| Body mass/kg | Surface area/$m^2$ | Heat loss/ $kJ\ hour^{-1}$ | Relative heat loss | |
|---|---|---|---|---|
| | | | $kJ\ hour^{-1}\ kg^{-1}$ | $kJ\ hour^{-1}\ m^{-2}$ |
| 31.2 | 1.08 | 191.1 | | |
| 24.0 | 0.88 | 171.0 | | |
| 18.2 | 0.77 | 146.4 | | |
| 9.6 | 0.53 | 108.9 | | |
| 3.2 | 0.24 | 49.1 | | |

Table 10.2

3 Copy this table and complete the two blank columns. The first of these columns shows the heat loss compared to the animal's body mass. The second shows the heat loss compared to the surface area of the animal.

*(2 marks)*

4 Which do you think is more important in determining heat loss, the animal's mass or its surface area? Explain the evidence from the table that supports your answer.

*(2 marks)*

5 Mammals produce the heat that results in a high body temperature from the metabolic reactions which take place in the body. Use the information in this assignment to explain why:

(a) when compared to their size, small mammals lose more body heat than large mammals;

(b) small mammals have a higher oxygen consumption relative to their body size than large mammals.

(c) small mammals benefit from haemoglobin that gives up its oxygen very readily.

*(5 marks)*

6 Since the early work on heat loss in dogs, studies have been carried out linking the body size of mammals with other factors. Explain the link between each of the following and your findings about the metabolic rates of different sized mammals:

(a) the number of mitochondria per gram of liver is higher in small mammals than in large mammals;

(b) the Etruscan shrew is the smallest of all mammals. Its mitochondria have many more cristae than the mitochondria of larger mammals;

(c) the tissues of small mammals have a higher density of capillaries than the tissues of large mammals.

*(5 marks)*

# Examination questions

1 The diagram shows an oxyhaemoglobin dissociation curve for human haemoglobin.

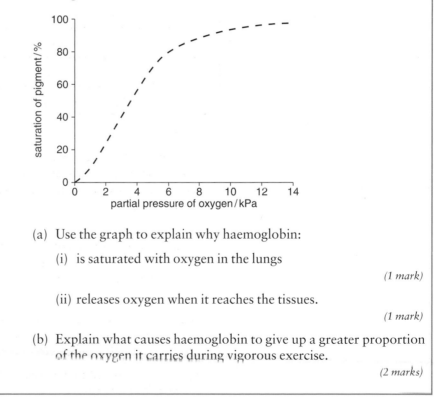

(a) Use the graph to explain why haemoglobin:

(i) is saturated with oxygen in the lungs

*(1 mark)*

(ii) releases oxygen when it reaches the tissues.

*(1 mark)*

(b) Explain what causes haemoglobin to give up a greater proportion of the oxygen it carries during vigorous exercise.

*(2 marks)*

# Nervous System

Parkinson's disease results when a group of nerve cells in the brain release insufficient amounts of a neurotransmitter, called dopamine. Normally, dopamine regulates the flow of nerve impulses that stimulate the contraction of specific skeletal muscles. When insufficient dopamine is released, these skeletal muscles contract irregularly, causing muscle tremors and involuntary movements. This is accompanied by stiffness in the joints.

Sufferers of Parkinson's disease can be treated with drugs that are converted to dopamine in the brain. In the 1980s, a more controversial treatment was discovered. This involved transplanting fetal brain cells into the brains of Parkinson's disease patients. The hope was that the transplanted fetal cells would mature in the recipient's brain and secrete dopamine. Most of the fetal cells that were used came from aborted fetuses. In Britain, fetal transplant treatment was stopped following a public outcry over this use of aborted fetuses. In the USA, the treatment continued.

In March 2001, the results of the first full clinical trial of the fetal transplant technique were published by a team based in Denver and New York. They discovered that the treatment did not help patients over the age of sixty. Whilst some younger patients did appear to benefit, the treatment produced effects that were worse than the disease itself in about 15% of recipients of the transplants. In these recipients, it appears that the fetal cells had grown too well and had begun to produce too much dopamine. What scientists now needed was a way to switch off the transplanted cells once they had done their work.

This story illustrates two aspects of medical research. Firstly, in countries such as Britain, democratic processes determine the use of advances in scientific techniques. Secondly, as science students, we should be cautious when reading about new 'wonder cures' until such time as full scientific trials have been completed.

## Transmission of impulses through the nervous system

In this part of the chapter, we will consider nervous transmission. In doing so, we will learn about the structure of nerve cells (**neurones**), the way that they transmit impulses along their length and the way they pass impulses from one to another.

## Neurones

Neurones are cells that are adapted to transmitting information through the nervous system. All neurones do this by:

- transmitting nerve impulses along their length

- stimulating other cells (target cells).

We need to be familiar with three types of neurone, each of which has a different function. Table 11.1 shows the names of these neurones and summarises their functions.

| Name of neurone | Stimulated by: | Transmits impulse to: |
| --- | --- | --- |
| Motor neurone (also called an effector neurone) | another neurone – either a relay neurone or a sensory neurone | an effector organ (gland or muscle) |
| Relay neurone | another relay neurone or a sensory neurone | another relay neurone or a motor neurone |
| Sensory neurone | a receptor, such as a Pacinian corpuscle, rod cell or cone cell, that is described in Chapter 12 | a relay neurone or a motor neurone |

**Table 11.1**
The three types of neurone that are found in the nervous system. In Chapter 12 you will see how sensory neurones are stimulated by receptors, such as Pacinian corpuscles, rod cells and cone cells.

**Q** 1 **Suggest why motor neurones are also called effector neurones.**

**Figure 11.1**
The structure of (a) a motor neurone, (b) a relay neurone and (c) a sensory neurone

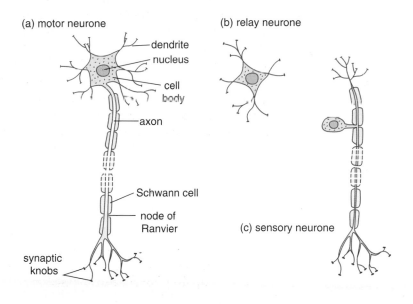

Figure 11.1 represents the structure of motor neurones, relay neurones and sensory neurones. The structure of a motor neurone is shown in greater detail than the other two types of neurone, reflecting the level of recall that is expected of you in the AQA Specification A. The motor neurone has a large **cell body**, which contains the nucleus and large numbers of cell organelles. The cell body has two types of cytoplasmic extensions that are adaptations to transmitting impulses.

- **Dendrons** are short extensions of the cell body. Each dendron has large numbers of smaller extensions, called **dendrites**. These dendrites are stimulated by other neurones and are the points at which impulse transmission always starts in a motor neurone. They transmit nerve impulses towards the cell body.

- The **axon** is a single extension, which can be up to 1 metre long in humans. The axon always transmits nerve impulses away from the cell body. The axon ends in a series of **synaptic knobs**, which are structures that stimulate the neurone's target organ (effector).

**Q** 2 **Which of the structures shown in the motor neurone is also present in relay neurones and sensory neurones?**

The motor neurone shown in Figure 11.1 has one further feature of interest to us: its axon is surrounded by **Schwann cells** with small gaps, called **nodes of Ranvier**, between them. As Figure 11.2 shows, each Schwann cell wraps itself around an axon. As a result, many layers of the surface membranes of Schwann cells surround each axon. This is important because the membrane of the Schwann cell is rich in a lipid, called **myelin**. Neurones that have Schwann cells wrapped around their dendrons and axons are called **myelinated neurones**. As we shall see in the next section of this chapter, myelinated neurones transmit impulses much faster than unmyelinated neurones.

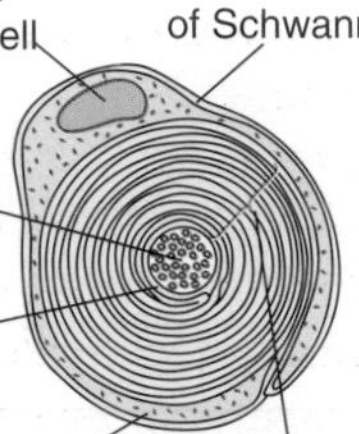

## Resting potentials in neurones

Charged ions are present in the cytoplasm of a cell and in the fluid surrounding a cell. As you learned during your AS course, the movement of ions across any membrane is restricted. The way in which membranes restrict the movement of potassium ($K^+$) ions and sodium ($Na^+$) ions, two ions that are important in nerve impulses, is summarised below and in Figure 11.3.

- The phospholipid bilayer prevents the diffusion of ions, such as potassium ($K^+$) ions and sodium ($Na^+$) ions.

- Intrinsic proteins, which span membranes, form **ion channels**. Some types of ion channel are permanently open, enabling constant diffusion of ions, such as $K^+$ and $Na^+$. Other types of channel have **voltage-sensitive 'gates'**, allowing diffusion only when they are opened. The surface membranes of neurones contain separate voltage-sensitive $K^+$ gates and voltage-sensitive $Na^+$ gates, which are important in controlling the transmission of impulses.

- Active transport of ions also occurs across membranes. One type of intrinsic protein, called a **sodium–potassium pump**, uses energy to remove $Na^+$ ions from the cytoplasm and take up $K^+$ ions into the cytoplasm.

nucleus of Schwann cell

surface membrane of Schwann cell

cytoplasm of axon of neurone

membrane of axon of neurone

cytoplasm of Schwann cell

layers of membrane where Schwann cell has wrapped itself around axon (= myelin sheath)

**Figure 11.2**
This cross-section through a myelinated axon shows how a Schwann cell has wrapped itself around the axon

**Figure 11.3**
The methods by which the movement of potassium and sodium ions across membranes are restricted

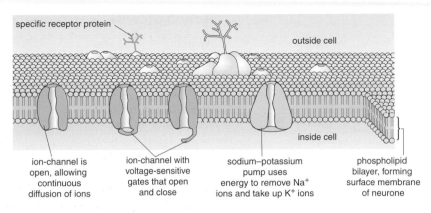

specific receptor protein

outside cell

inside cell

| ion-channel is open, allowing continuous diffusion of ions | ion-channel with voltage-sensitive gates that open and close | sodium–potassium pump uses energy to remove Na⁺ ions and take up K⁺ ions | phospholipid bilayer, forming surface membrane of neurone |

Table 11.2 shows the concentration of three types of ion inside and outside a mammalian motor neurone when it is not transmitting an impulse ('resting' neurone). As we have seen above, $Na^+$ ions and $K^+$ ions can move across the surface membrane of the neurone through open ion channels. However, the movement of each type of ion is determined by its electrochemical gradient, which is made up of both the electrical gradient and concentration gradient of that ion across the membrane. Take the $K^+$ ions in Table 11.2 for example. You might expect $K^+$ ions to move out of the neurone down their concentration gradient. However, the build-up of positive charges ($Na^+$ ions and $K^+$ ions) outside the membrane eventually repels the movement of any more $K^+$ ions, preventing their outward movement. An equilibrium is reached when these effects balance each other out and no net movement of $K^+$ ions occurs. Table 11.2 also shows a concentration gradient of $Cl^-$ ions, from which you might expect $Cl^-$ ions to move into the neurone. However, the cytoplasm contains large, negatively charged protein molecules that cannot cross the surface membrane and which repel incoming $Cl^-$ ions, preventing their movement into the cell. The imbalance of ions shown in Table 11.2 causes a potential difference (or voltage) between the inside of the neurone and its surroundings. Since a neurone in this condition is not transmitting an impulse, this potential difference is called a **resting potential**. The value of this resting potential is different in different types of neurone, but in the mammalian motor neurone in Table 11.2 it is −70 mV. A membrane that has different charges on its inside and outside is **polarised**.

| Ion | Concentration inside cell/mmol dm$^{-3}$ | Concentration outside cell/mmol dm$^{-3}$ |
|---|---|---|
| $K^+$ | 150.0 | 2.5 |
| $Na^+$ | 15.0 | 145.0 |
| $Cl^-$ | 9.0 | 101.0 |

**Table 11.2**
The concentration of three types of ion inside and outside a 'resting' mammalian motor neurone. The concentrations shown in the table result in a resting membrane potential of −70 mV

**Q 3 Why is the membrane of a neurone said to be polarised?**

## Action potentials in neurones

An **action potential** is an abrupt but short-lived reversal of the resting potential of a neurone. It occurs at a specific point of the neurone – we will see how a wave of action potentials passes along a neurone later in this chapter. Figure 11.4 shows the electrical changes that occur during a single action potential.

1  The action potential begins when a particular stimulus causes the membrane at one part of the neurone suddenly to increase its permeability to Na$^+$ ions. This is labelled **A** in Figure 11.4 and happens because the stimulus has caused voltage-sensitive Na$^+$ gates in this part of the membrane to open. As a result, Na$^+$ ions diffuse rapidly into the neurone along their electrochemical gradient, reducing the negativity inside that part of the neurone. In a positive feedback loop, this causes more and more voltage-sensitive Na$^+$ gates to open until the voltage difference across the membrane reverses. The highest positive membrane potential is the action potential and this part of the membrane is now said to be **depolarised**.

2  At a certain point, labelled **B** in Figure 11.4, the depolarisation of the membrane causes the voltage-sensitive Na$^+$ gates to close.

3  Shortly after the voltage-sensitive Na$^+$ gates close, the voltage-sensitive K$^+$ gates open. This is labelled **C** in Figure 11.4. The voltage-sensitive K$^+$ gates open more slowly than do the voltage-sensitive Na$^+$ gates and allow K$^+$ ions to flow out of the neurone, along their electrochemical gradient. This outflow of K$^+$ ions restores the resting potential in the neurone. In a negative feedback loop, the outward movement of K$^+$ ions results in the closing of the voltage-sensitive K$^+$ gates (labelled **D** in Figure 11.4). There is a slight 'overshoot' in the movement of K$^+$ ions, which causes the membrane potential to become slightly lower than its normal resting potential. This is called **hyperpolarisation** and is labelled region **E** in Figure 11.4.

4  After the resting potential has been restored, sodium-potassium pumps move Na$^+$ ions out of the neurone and move K$^+$ ions back into the neurone by active transport.

**Figure 11.4**
The changes in membrane potential associated with an action potential at one tiny part of the surface membrane of a neurone. Voltage-sensitive Na$^+$ gates open at point **A** and close at point **B**. Voltage-sensitive K$^+$ gates open at point **C** and close at point **D**

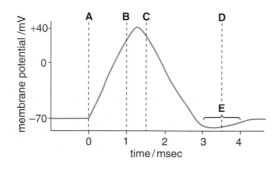

**Q**  4  Use Figure 11.4 to give the value of:
(a) the resting potential
(b) the action potential
(c) the length of time for which the action potential lasts.

**Extension box 1**

## All or nothing principle

To cause an action potential, a stimulus must cause sufficient movement of $Na^+$ ions and $K^+$ ions to depolarise the membrane. An impulse that is strong enough to cause an action potential is called a **threshold stimulus**. Figure 11.5 represents the threshold stimulus. If a stimulus is as strong as, or stronger than, the threshold, a nerve impulse is transmitted along the entire neurone at a constant and maximum strength. The transmission of the impulse is independent of any further intensity of the stimulus. This is the **all or nothing principle**: a stimulated neurone either fails to transmit an impulse at all or it transmits an impulse at a constant strength along its whole length.

**Figure 11.5**
An action potential has been redrawn to show the threshold and the refractory period. The threshold shows the minimum disturbance in membrane potential that must be caused by a stimulus to produce an action potential. The refractory period represents a time during which the membrane cannot be depolarised again

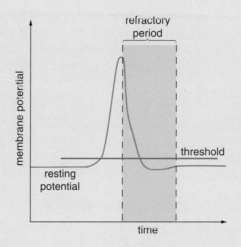

A stimulus that is weaker than a threshold stimulus is called a **subthreshold stimulus**. It is incapable of causing an action potential. However, if a second stimulus, or a series of subthreshold stimuli, is quickly applied to the neurone, the cumulative effect might be enough to cause an action potential. This is called **summation**. You will learn in Chapter 12 how rod cells work together in groups to produce a threshold stimulus on a single relay cell, thereby helping us to see in dim light conditions.

## Conduction of a nerve impulse along a single neurone

The above account explains what happens at a tiny part of the surface membrane of a neurone as it becomes depolarised. We now need to understand how this depolarisation enables a neurone to transmit an impulse all the way along its length towards its target cell.

1　Look at Figure 11.6(a). This represents a membrane in which all the voltage-sensitive $Na^+$ gates are closed. The entire membrane shown is polarised and the length of neurone shown has a resting potential.

2　In Figure 11.6(b) voltage-sensitive $Na^+$ gates have opened and the resulting influx of sodium ions has caused an action potential.

3 Figure 11.6(c) shows how the action potential has caused the voltage-sensitive $Na^+$ gates to close and voltage-sensitive $K^+$ gates to open. Potassium ions flow out of the neurone, restoring the resting potential. However, the action potential has disturbed the adjacent membrane causing its voltage-sensitive $Na^+$ gates to open. A new action potential is occurring in this part of the neurone membrane.

4 In Figure 11.6(d) the action of Figure 11.6(c) has been repeated and a third action potential has been produced. Although the resting potential has been restored, the balance of $Na^+$ ions and $K^+$ ions has not. This is done by sodium–potassium pumps using the energy released by the breakdown of ATP. This is the only time that active transport is involved in this entire process.

In this way, a single action potential results in a step-by-step wave of action potentials along the entire length of the membrane. This is how a nerve impulse travels along a neurone.

**Figure 11.6**
Conduction of a nerve impulse along an unmyelinated neurone.
(a) The neurone is at rest.
(b) A tiny part of the membrane on the left of the diagram has become depolarised and an action potential has occurred.
(c) This action potential has depolarised another tiny part of the membrane immediately adjacent to it so that a second action potential occurs.
(d) The second action has depolarised the next tiny section of membrane and a third action potential has occurred. In this way, the action potential moves steadily from left to right of the diagram

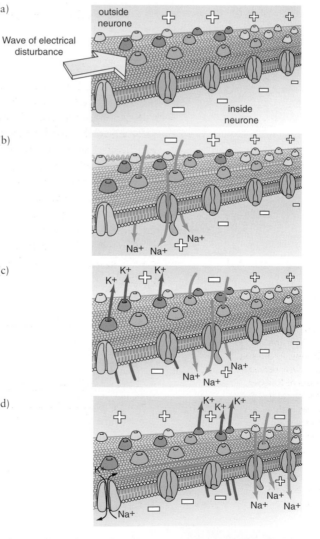

(a)

(b)

(c)

(d)

**Q** 5 What is the colour of sodium–potassium pumps in Figure 11.6?

Whilst following Figure 11.6(d), you might have wondered why the second action potential depolarised only the membrane to its right and did not depolarise the membrane to its left as well. This is because the membrane to its left was recovering from the previous action potential in a short time, called the **refractory period**. If you look back to Figure 11.5, you will see this refractory period represented during an action potential. During the refractory period of a section of membrane, its voltage-sensitive Na⁺ gates cannot be opened. As a result, the membrane immediately 'behind' an action potential cannot be depolarised and so cannot set up a further action potential of its own. This has two important consequences. Firstly, it explains why nerve impulses can only travel in one direction along a neurone. An action potential can only depolarise the membrane 'in front' as the membrane 'behind' is in its refractory period and cannot be depolarised again. Secondly, because the refractory period lasts for up to 10 milliseconds, it limits the frequency with which neurones can transmit impulses.

**Q** 6 Why do impulses only travel in one direction along a neurone?

## Speed of conduction of impulses

The above account of conduction of a nerve impulse is true for unmyelinated neurones. Each action potential triggers the next one in a self-propagating wave. However, conduction of nerve impulses is affected by the myelin in the membranes of Schwann cells that surround myelinated neurones. Myelin almost stops the diffusion of Na⁺ ions and K⁺ ions. In other words, it insulates those parts of the neurone that it covers. As a result, the diffusion of Na⁺ ions and K⁺ ions can only occur at the nodes of Ranvier, shown in Figure 11.1. If you look at Figure 11.7 you will see the consequence of this. Figure 11.7 shows that action potentials 'jump' from one node of Ranvier to the next. This is known as **saltatory conduction** (from the Latin *saltare*, meaning to jump) and results in impulses travelling much faster in myelinated neurones than in unmyelinated neurones.

**Figure 11.7**
Myelin in the surface membranes of Schwann cells prevents the diffusion of Na⁺ ions and K⁺ ions. As a result, local currents flow between adjacent nodes of Ranvier, shown by the circular arrows in the drawing. This causes depolarisation only at the nodes of Ranvier, so that action potentials 'jump' from one node of Ranvier to the next. This speeds up the rate of transmission along myelinated neurones

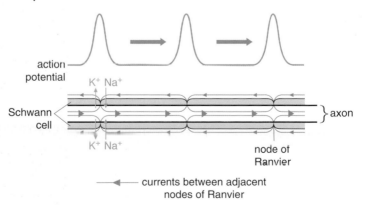

Irrespective of whether they are myelinated or unmyelinated, temperature affects the speed at which neurones conduct impulses. As you learned in your AS Biology course, temperature affects the rate of diffusion. Provided they do not begin to denature cell proteins, high temperatures cause an increase in the rate of diffusion of ions. As a result, they will increase the rate of conduction of nerve impulses. Impulses are also faster in an axon with a large diameter. This is because there is less 'leakage' of ions in neurones with a large diameter than in neurones of small diameter. Since ion leakage weakens membrane potentials, the conduction of action potentials is often lost along unmyelinated neurones with a small diameter. As we have already seen, the presence of myelin stops ion leakage, so diameter is only an important consideration for unmyelinated neurones.

**Q** 7 **Use your understanding of the effects of size to suggest why leakage of ions is greater in small neurones than in larger neurones.**

## Transmission at synapses

Nervous pathways involve chains of at least two neurones that pass impulses from one to another. The junction between two neurones is called a **synapse**. If you look at Figure 11.8 you will see that the two neurones at a synapse do not touch each other. Instead, there is a gap of about 20 µm between one neurone and the next in the chain. This gap is called a **synaptic cleft**. On one side of the synaptic cleft is the **presynaptic membrane**, which is the surface membrane at the end of one neurone (the presynaptic neurone). On the other side of the synaptic cleft is the **postsynaptic membrane**, which is the surface membrane of the next neurone in the chain (the postsynaptic neurone). When a nerve impulse arrives at the presynaptic membrane of a synapse it causes the release of a chemical **neurotransmitter**. Molecules of this neurotransmitter diffuse across the synaptic cleft and cause depolarisation in the surface membrane of the postsynaptic neurone.

Figure 11.8 shows an **excitatory synapse**, i.e. one that causes an action potential in the postsynaptic cell. Notice that the end of the presynaptic neurone forms a swelling, called the **synaptic knob**, which contains small vesicles containing neurotransmitter. When an impulse arrives at the synaptic knob, it causes gated calcium ion channels in the presynaptic membrane to open. As a result, calcium ($Ca^{2+}$) ions flow into the synaptic knob along their electrochemical gradient, causing the vesicles to move to the presynaptic membrane and fuse with it. Molecules of the neurotransmitter are then released from the vesicles and diffuse across the synaptic cleft. They bind to specific receptor proteins in the postsynaptic membrane. These receptor proteins are attached to gated ion channels and change their shape when they bind with molecules of neurotransmitter. As a result, ions diffuse across the postsynaptic membrane. If the diffusion of ions reaches a threshold value, it causes an action potential in the postsynaptic membrane. Once started, the action potential is conducted along the postsynaptic neurone in the same way as shown in Figure 11.6. Figure 11.9 summarises the events that occur during transmission of an impulse across a synapse.

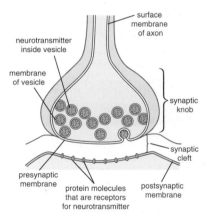

neurotransmitter inside vesicle

membrane of vesicle

surface membrane of axon

synaptic knob

synaptic cleft

presynaptic membrane

protein molecules that are receptors for neurotransmitter

postsynaptic membrane

**Figure 11.8**
The structure of an excitatory synapse. When an impulse arrives at the end of a presynaptic neurone it causes vesicles of neurotransmitter to fuse with the presynaptic membrane and diffuse through the synaptic cleft to the postsynaptic membrane. Here they bind with specific receptors and cause depolarisation of the postsynaptic membrane

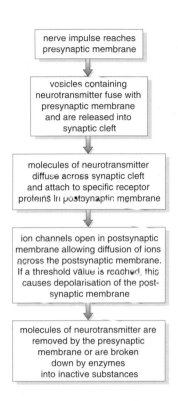

**Figure 11.9**
The flowchart summarises transmission of a nerve impulse across an excitatory synapse

**Q** 8 Use your knowledge of AS Human Biology to name the process by which neurotransmitter is released from the vesicles in the synaptic knob.

The action of neurotransmitters is short-lived because molecules of neurotransmitters are quickly removed from the postsynaptic membrane. This happens because:

- molecules of neurotransmitter diffuse out of the synaptic cleft

- molecules of neurotransmitter are taken up by the membrane of the presynaptic membrane by endocytosis

- enzymes quickly break down molecules of neurotransmitter into inactive substances.

There are many different neurotransmitters in the nervous system. You need to know only two. **Acetylcholine** is released by motor neurones on to muscle cells and by neurones in the parasympathetic division of the autonomic nervous system. **Noradrenaline** is released by neurones in the sympathetic division of the autonomic nervous system. However, many substances are similar in shape to natural neurotransmitters and can fit into the specific protein receptors of postsynaptic membranes. **Antagonistic substances** bring about the same effect as the neurotransmitter when they bind to specific protein receptors of postsynaptic membranes. Anatoxin is an example of an antagonistic

| Aspect of synaptic transmission | Explanation |
| --- | --- |
| Unidirectionality | Because neurotransmitters are released only by the presynaptic membrane, transmission at a synapse can only flow in the direction: presynaptic neurone → postsynaptic neurone. |
| Summation | Sometimes insufficient neurotransmitter is released from a single presynaptic neurone to reach the threshold needed to depolarise the membrane of the postsynaptic membrane. In this case, simultaneous release of neurotransmitter from several presynaptic neurones is needed to cause depolarisation of the postsynaptic membrane. Groups of rod cells do this when they synapse with individual relay neurones in the retina (see Chapter 12). |
| Inhibition | In some synapses, movement of ions causes the inside of the postsynaptic neurone to become more negative than its resting potential. This is called **hyperpolarisation** and inhibits stimulation of the postsynaptic neurone. Synapses like this are called **inhibitory synapses**. The AQA Specification A does not require you to know the mechanism of transmission by inhibitory synapses, but we will come across them again when we learn about the action of rod cells in Chapter 12. |

**Table 11.3**
Unidirectionality, summation and inhibition in synapses

235

substance. It is produced by some algae and mimics the effects of acetylcholine. Since anatoxin is swallowed in contaminated water, its effects are strongest in the mouth, where anatoxin causes continuous salivation. **Antagonistic substances** also bind to specific protein receptor sites on postsynaptic membranes. Once this has happened, antagonistic substances prevent neurotransmitter molecules from binding with their receptor sites and so block the action of the neurotransmitters. High blood pressure can be treated using drugs called β-**blockers**. These drugs are antagonists of adrenaline-receptors on the surface membrane of muscle cells in the heart. Curare is another type of receptor antagonist, which blocks the action of acetylcholine at the junction of nerves and muscles. This makes curare useful as a general muscle relaxant in patients undergoing major surgery.

Table 11.3 summarises some aspects of synaptic transmission.

## The spinal cord and spinal reflexes

In Chapter 12, we will learn how some changes in the internal or external environment (called **stimuli**) cause impulses to travel along sensory neurones. Here we will learn how impulses travelling along sensory neurones result in a change in behaviour. The simplest behaviour patterns are called reflexes. A **simple reflex** is in-born (i.e. is not learned) and always results in the same, fixed response to a particular stimulus. Examples include dilating your pupils in dim light, producing saliva when you taste food and withdrawing part of your body from a painful stimulus. The nervous pathway of a simple reflex is called a **reflex arc**. Sometimes a reflex arc involves parts of the brain, e.g. those that control the muscles in the iris and tension in the suspensory ligaments of the eye, which you will learn about in Chapter 12. Reflex arcs that involve the spinal cord, but not the brain, are called **spinal reflexes**.

**Figure 11.10**
Cross-section through part of a human spinal cord. The grey matter contains unmyelinated relay cells. The white matter contains myelinated sensory and motor neurones. The position of three neurones involved in a spinal reflex is shown. Motor neurones always leave the spinal cord via the ventral root. Sensory neurones always enter the spinal cord via the dorsal root and have their cell bodies in the dorsal root ganglion

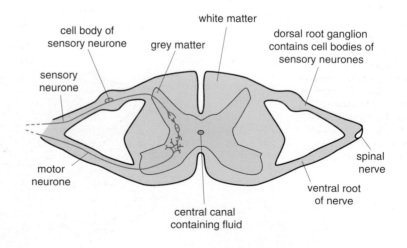

**Q** 9 Give *two* characteristics of a simple reflex.

The human **spinal cord** is a hollow tube of nervous tissue that runs from the brain, through a large space in the backbone, to the base of the spine. Thirty-one pairs of spinal nerves enter and leave the spinal cord along its length. They all contain large numbers of sensory and motor neurones. A region within the spinal cord, called the **grey matter**, contains large numbers of relay neurones. Figure 11.10 represents part of the spinal cord and one spinal nerve. It also shows three neurones involved in a reflex arc.

Imagine you have touched the hot plate of an iron whilst ironing your jeans. A simple reflex would occur in response to this stimulus – you would quickly pull your hand away from the iron. In this example, temperature receptors and pain receptors in your skin detect the stimulus. As a result, they produce generator potentials and pass a nerve impulse to sensory neurones just under your skin. (You will learn about generator potentials in Chapter 12.) These sensory neurones carry their impulses right along their length and into the spinal cord, via a spinal nerve. Sensory neurones always enter the spinal cord via the **dorsal root** of the nerve. (You would normally call your dorsal surface the 'back' of your body.) The swelling in the dorsal root that contains the cell bodies of all these sensory neurones is called the **dorsal root ganglion**. In the spinal cord, sensory neurones synapse with relay neurones that, in turn, synapse with motor neurones. All these synapses occur in the **grey matter** of the spinal cord. The motor neurones that carry impulses to muscle cells in your biceps muscle leave the spinal cord via the **ventral root** of the spinal nerve. (You would normally call your ventral surface the 'front' of your body.) Eventually, the motor cells release neurotransmitter onto muscle cells in your biceps, causing them to contract. As a result, you bend your arm and pull your hand away from the hot iron. This action is rapid and automatic – you did not have to think about it. In reality, large numbers of receptors, sensory neurones, relay neurones and motor neurones would be involved. For the sake of simplicity, only one of each type of neurone is shown in Figure 11.10.

**Figure 11.11**
This cross-section through the spinal cord shows how relay neurones run up ascending tracts and down descending tracts in the grey matter. These relay neurones enable us to be aware of stimuli that have caused a simple reflex. They also enable us consciously to modify simple reflexes

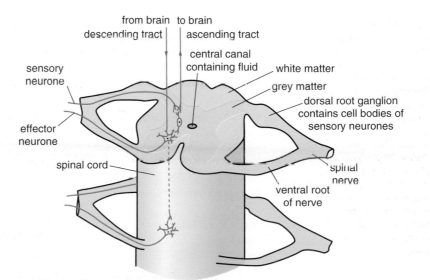

The withdrawal reflex above is obviously advantageous. By withdrawing your hand from a damaging stimulus, you protect your body. Luckily for us, the reflex arcs are there at birth – we do not have to learn to withdraw from harmful stimuli the hard way. However, we can learn to modify these reflexes. For example, in addition to withdrawing your hand from a hot iron, you might shout something like 'Ouch!'. This is a learned response and is controlled by nervous pathways that you developed during the learning process. Figure 11.11 shows how this is controlled. In this case, additional relay neurones carry impulses from the reflex arc up **ascending tracts** in the grey matter of the spinal cord to your brain. In your brain, new pathways of relay neurones are formed that eventually stimulate motor neurones to the muscles that control speech. Thus, simple reflexes can sometimes be adapted through learning.

**Q 10** Imagine you have spent several hours preparing a delicious meal. When you take the meal from the oven, the container burns your hand. Suggest the nervous pathway that prevents you from dropping the meal on the kitchen floor.

## The autonomic nervous system

The **autonomic nervous system** is a part of the nervous system that controls internal glands and muscles that are normally beyond your conscious control (Table 11.4). The autonomic nervous system has two divisions.

Table 11.4
The effects of stimulation by the parasympathetic and sympathetic divisions of the autonomic nervous system are usually antagonistic

- The **parasympathetic division** of the autonomic nervous system generally has an inhibitory effect and helps relaxation. The motor neurones of the parasympathetic division release the neurotransmitter acetycholine on to the glands or organs that they stimulate.

| Target organ or tissue | Effect of parasympathetic stimulation | Effect of sympathetic stimulation |
|---|---|---|
| Iris of eye | constricts pupil | dilates pupil |
| Salivary gland | stimulates secretion of saliva | inhibits secretion of saliva |
| Intercostal muscles | decreases breathing rate | increases breathing rate |
| Bronchi and bronchioles | constricts these tubes | dilates these tubes |
| Blood vessels | dilates blood vessels – decreasing blood pressure within them | constricts blood vessels – increasing blood pressure within them |
| Heart | decreases heart rate and stroke volume | increases heart rate and stroke volume |
| Gut | stimulates peristalsis | inhibits peristalsis |
| Sweat glands | no effect | increases sweat production |

● The **sympathetic division** of the autonomic nervous system generally has an excitatory effect and helps the body to react to stress. The motor neurones of the sympathetic division release the neurotransmitter noradrenaline on to the glands or organs that they stimulate.

**Extension box 2**

## Control of the heart rate

You learned about the structure and action of the heart in your AS Human Biology course. You learned that heart muscle is myogenic, i.e. it contracts of its own accord. You also learned how the activity of the sinoatrial node (SA node), atrioventricular node (AV node) and Purkyne tissue coordinate the contraction of heart muscle to produce a heart beat. The autonomic nervous system is able to change the rate of contraction of the heart.

Figure 11.12 shows parts of the autonomic nervous system involved in controlling the heart beat. A part of the brain, called the **medulla**, contains two regions that control the rate at which the heart beats. The **cardioacceleratory centre (CAC)** speeds up the heart rate. Sympathetic neurones run from the CAC down a descending tract of the spinal cord and along a spinal nerve to the sinoatrial node, atrioventricular node and parts of the heart muscle itself. When the CAC is stimulated, nerve impulses travel along these sympathetic neurones and they release noradrenaline on to their target cells. This causes the rate of heartbeat to increase.

**Figure 11.12**
The autonomic nervous system controls the rate at which the heart beats. Pressure receptors in the walls of the aorta and carotid artery stimulate the CAC and CIC in the medulla of the brain. Sympathetic neurones from the CAC stimulate an increase in heart rate; parasympathetic neurones from the CIC stimulate a decrease in the heart rate

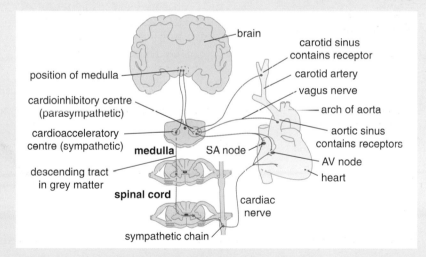

**Q 11** Name the antagonist, commonly used as a drug, which blocks the action of the cardioacceleratory centre on the heart.

The medulla contains a second group of neurones, called the **cardioinhibitory centre (CIC)**. Parasympathetic neurones from the CIC leave the brain in the vagus nerve and run to the sinoatrial node and atrioventricular node. When the CIC is stimulated, nerve impulses travel along these parasympathetic neurones and they release acetycholine on to their target cells. This causes the rate of heartbeat to decrease.

In this way, opposing stimulatory sympathetic neurones and inhibitory parasympathetic neurones control the rate of heartbeat. They do so through simple reflexes in response to blood pressure. The reflexes are started by pressure receptors in the wall of the aorta and the wall of the carotid artery (passing from the aorta to the head), which send nerve impulses to the CAC and CIC in the medulla.

# Summary

- The nervous system is made of three types of neurone: sensory neurones, relay neurones and motor neurones. Sensory and motor neurones can be myelinated, in which case they transmit impulses faster than unmyelinated neurones.

- A neurone has a resting potential, which is a voltage across its surface membrane. This results from an imbalance in ions, especially $Na^+$ and $K^+$ ions across its membrane. When an impulse is passed along an excitatory neurone, this resting potential is reversed, producing an action potential.

- The transmission of a nerve impulse along a neurone depends on the wave-like propagation of action potentials along its length. In myelinated neurones, the action potentials 'jump' from one node of Ranvier to the next.

- Because the surface membrane of a neurone has a refractory period, impulses can only pass in one direction along a neurone. The refractory period also limits the frequency with which neurones can pass separate impulses.

- So long as a threshold level of stimulation is reached, a neurone transmits an impulse along its length at a constant and maximum strength. This is the 'all or nothing' principle.

- Nerve impulses pass from neurone to neurone across small gaps, called synapses. These are bridged by neurotransmitters that are secreted by small synaptic knobs at the end of neurones. Acetycholine and noradrenaline are examples of neurotransmitters.

- The spinal cord runs from the brain to the base of the spine. In a simple spinal reflex arc sensory neurones synapse with relay neurones and relay neurones synapse with motor neurones within the grey matter of the spinal cord. These simple spinal reflexes protect the body from damage but can be modified as a result of learning.

- The autonomic nervous system controls internal glands and muscles over which we normally have little conscious control. The parasympathetic division of the autonomic nervous system secretes acetylcholine and helps relaxation whereas the sympathetic division of the autonomic nervous system secretes noradrenaline and helps in preparation for emergencies.

- Changes in the heart rate are controlled by the balancing action of the parasympathetic and sympathetic divisions of the autonomic nervous system.

# Assignment

## Snakes, puffer fish and paralysis

You may not have a lot of sympathy for poisonous snakes – not many people have! But think for a minute about the problems that such a snake faces in catching and swallowing a large rat. Rats move fast but snakes are relatively slow. So it is a question of first stop your rat. In addition, a snake cannot chew its prey or bite pieces off. This means that once caught, the rat must be swallowed whole. Clearly a rat must be subdued before it is swallowed or the snake could come to considerable harm. Observations such as these suggested that snakes and other poisonous animals produce substances in their venom which in some way affect the nerves and muscles of their prey, often bringing about paralysis.

In this assignment we will look in more detail at the way in which substances produced by poisonous animals affect our nerves and muscles. You will need to use material from various parts of the specification as well as from this chapter to answer the questions. Before you start, it would be worth looking up the following key topics either in your notes or in your textbook:

● proteins and protein structure

● plasma membranes

● diffusion and active transport.

Look at Figure 11.13; it shows a voltage-sensitive sodium gate in the axon membrane.

1 (a) Describe the part played by this voltage-sensitive sodium gate in producing an action potential.

*(2 marks)*

(b) The plasma membranes of nerve cells contain many other proteins like the one shown in the diagram. Some are sensitive to sodium ions, some to potassium ions and some to calcium ions. Use the diagram to suggest why each of these proteins is specific to a particular ion.

*(2 marks)*

2 (a) The puffer fish is regarded as a great delicacy in Japan. The restaurants that serve it, however, have to be specially licensed. This is because some of the organs in the body contain an extremely poisonous substance, tetrodotoxin. Unless the fish is prepared properly, you are very likely to be poisoned. Tetrodotoxin locks into sites at the entrance to the ion channel in the sodium gate. Use this information to suggest why one of the symptoms of being poisoned by tetrodotoxin is paralysis.

*(2 marks)*

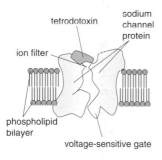

**Figure 11.13**
A voltage-sensitive sodium gate is one of a number of different sorts of protein in the plasma membrane surrounding an axon

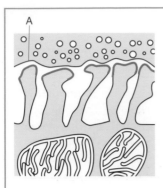

**Figure 11.14**
A neuromuscular junction is very similar to a synapse. This drawing has been taken from an electronmicrograph. It has been injected with bungaratoxin and then stained with bungaratoxin antibodies. In this drawing the position of the bungaratoxin antibodies are shown in red

We will have a look now at a neuromuscular junction. This is where a neurone meets a muscle. Neuromuscular junctions are very similar to synapses between two neurones, the only real difference is that transmission across a neuromuscular junction will lead to a muscle contracting rather than another nerve impulse. Figure 11.14 shows a neuromuscular junction.

3  Look at Figure 11.14. What is the evidence from this drawing that the feature labelled **A** is the presynaptic membrane?

*(1 mark)*

The krait is a small but extremely poisonous snake found in South-east Asia. In 1963 a small polypeptide consisting of approximately 70 amino acids was isolated from krait venom. This substance was called bungaratoxin, and was found to cause paralysis when very small amounts were injected into mice. When the tissues of these injected mice were treated with fluorescent antibodies and examined with a microscope, it was found that the toxin occurred only at the neuromuscular junction.

4  Explain why bungaratoxin only has an effect if it is injected into an animal. It has no effect if it is simply swallowed.

*(2 marks)*

5  (a)  Use your knowledge of antibodies to explain how fluorescent antibodies can be used to locate the position of bungaratoxin in the neuromuscular junction.

*(3 marks)*

(b)  Use the information given in this question so far to suggest why animals bitten by a krait often die because the muscles associated with breathing become paralysed.

*(4 marks)*

6  Table 11.5 shows a number of other substances and describes briefly the way in which they affect the nervous system. For each substance, describe how the effect could lead to the consequence described in the last column of the table.

*(4 marks)*

**Table 11.5**  The effects of a variety of poisons on the nervous system

| Substance | Effect | Consequence |
|-----------|--------|-------------|
| Black widow spider venom | Causes large amounts of acetylcholine to be released | After a short period of time synaptic transmission stops |
| Organophosphate insecticides | Inhibit the enzyme acetylcholinesterase | One of the symptoms of accidental poisoning is the production of large quantities of tears and saliva |
| W-conotoxin produced by cone shells | Prevents $Ca^{2+}$ ions from crossing the pre-synaptic membrane | Paralysis |

# Examination questions

1  The diagram shows some of the pump and channel proteins in the cell surface membrane of a neurone.

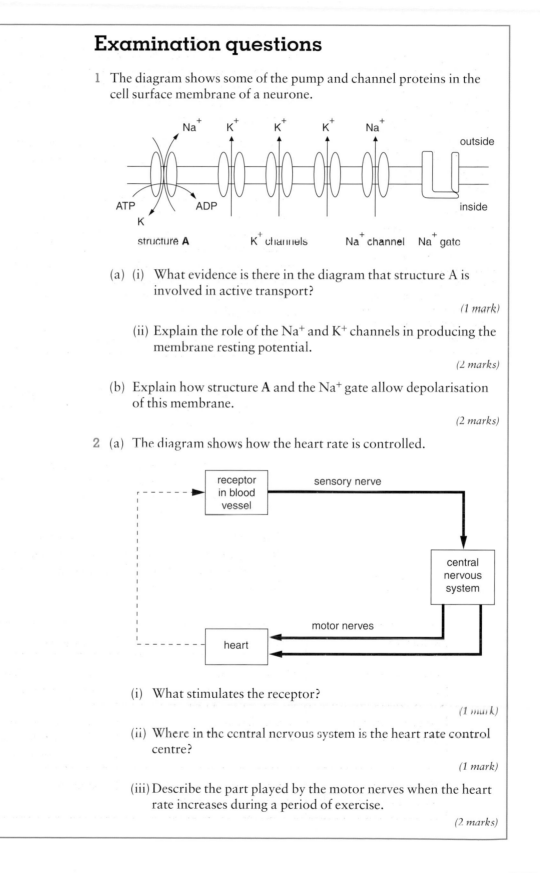

(a) (i)  What evidence is there in the diagram that structure A is involved in active transport?

*(1 mark)*

(ii)  Explain the role of the $Na^+$ and $K^+$ channels in producing the membrane resting potential.

*(2 marks)*

(b)  Explain how structure **A** and the $Na^+$ gate allow depolarisation of this membrane.

*(2 marks)*

2  (a)  The diagram shows how the heart rate is controlled.

(i)  What stimulates the receptor?

*(1 mark)*

(ii)  Where in the central nervous system is the heart rate control centre?

*(1 mark)*

(iii) Describe the part played by the motor nerves when the heart rate increases during a period of exercise.

*(2 marks)*

(iv) Use the diagram to explain how negative feedback would operate in the control of heart rate.

*(2 marks)*

3   The diagram represents a reflex arc.

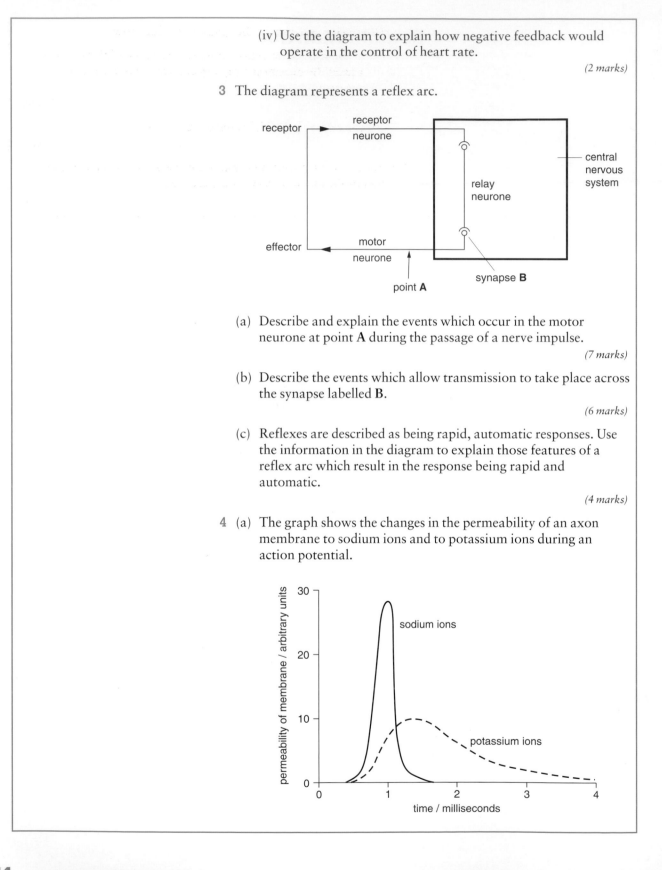

(a)   Describe and explain the events which occur in the motor neurone at point **A** during the passage of a nerve impulse.

*(7 marks)*

(b)   Describe the events which allow transmission to take place across the synapse labelled **B**.

*(6 marks)*

(c)   Reflexes are described as being rapid, automatic responses. Use the information in the diagram to explain those features of a reflex arc which result in the response being rapid and automatic.

*(4 marks)*

4   (a)   The graph shows the changes in the permeability of an axon membrane to sodium ions and to potassium ions during an action potential.

Use the information in the graph to explain how:

(i) at the start of an action potential, the potential difference across the membrane rapidly changes from negative to positive;

*(2 marks)*

(ii) the resting potential is restored.

*(1 mark)*

(b) Suggest why, during a period of intense nervous activity, the metabolic rate of a nerve cell increases.

*(2 marks)*

# Receptors and Effectors

People often say that 'seeing is believing'. However, can we always trust our own eyes? Figure 12.1 shows a black spot and a red cross. Hold this book at arm's length and close your left eye or cover it with your free hand. Look directly at the black spot, using only your right eye. You will be able to see both the black spot and the red cross.

**Figure 12.1**
The black spot and red cross can be used to cause optical illusions. Follow the instructions in the Introduction to see two ways that you can 'fool' your brain

Now slowly bring the book closer to your face. Look at the spot all the time you do this. You will find that, at a certain distance from your face, you will still see both the spot and the cross but that both will appear black. When the book is a little nearer your face, you will still see the black spot but the cross will 'disappear' altogether. At this distance, the page to the right of the spot will appear white.

These are examples of optical illusions. Moving the book makes the images of the spot and cross fall on different parts of the retina inside your right eye. When you see both the black spot and the red cross, their images are falling on cone cells in the retina of your right eye. Cone cells detect shape as well as colour. When the red cross appears to be black, its image is falling on rod cells in the retina of your right eye. Rod cells detect shape but cannot detect colour, so you see the cross but it appears to be black. When the red cross 'disappears', its image has fallen on the blind spot in your right retina. Since this blind spot has no light-sensitive cells, no image is transmitted to your brain. The part of your brain that interprets images on your retina cannot cope with no image, so it fills in the 'blind area' with what is immediately around it, i.e. the whiteness of the rest of the page.

In this chapter, you will learn more about the structure of the eye and about the way that cells in the retina convert images into nerve impulses that pass to your brain.

**Q** 1 Suggest why you are normally unaware of a 'blind area' in each eye.

## Receptors

A **stimulus** is a change in the environment that is capable of causing a response by the nervous system. Very few neurones are directly sensitive to stimuli. This is where receptors come in. A **receptor** is a transducer. This means that it converts a stimulus into a nerve impulse. As Figure 12.2 shows, a stimulus causes a change in the membrane potential of the receptor, resulting in a **generator potential** in that receptor. In turn, this generator potential results in an action potential along a sensory neurone.

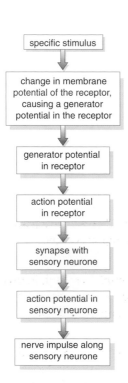

**Figure 12.2**
A flowchart showing the stages by which a receptor converts a stimulus into a nerve impulse

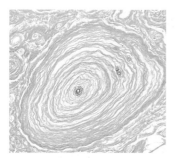

**Figure 12.3**
A light micrograph of a single Pacinian corpuscle in longitudinal section. The central neurone is surrounded by layers of connective tissue, making it look like a sliced onion

Receptors are quite specific in the stimuli that will cause generator potentials. We need to be familiar with only two receptors: Pacinian corpuscles and the receptor cells in the retina of the eye. Each responds to only one type of stimulus: all other stimuli fail to cause a generator potential in these receptors.

## Pacinian corpuscles

Pacinian corpuscles are found beneath the skin, around joints and tendons, in the external genitalia of both sexes and in some internal organs. They are sensitive to changes in mechanical pressure. For example, Pacinian corpuscles in the skin of your feet enable you to detect pressure changes when you put on a pair of trainers. Figure 12.3 shows a light micrograph of a single Pacinian corpuscle in longitudinal section. It resembles a sliced onion, containing layers of connective tissue around the umyelinated end of a myelinated neurone.

We have seen the importance of ion channels in causing membrane potentials. The neurone inside a Pacinian corpuscle has **stretch-mediated $Na^+$ channels** in its surface membrane. These sodium channels are so called because they change their shape when stimulated by changes in pressure, so that their permeability to $Na^+$ ions changes.

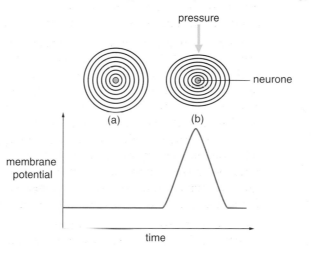

**Figure 12.4**
(a) In an unstimulated Pacinian corpuscle, the layers of connective tissue and the neurone that they enclose have a rounded cross-section.
(b) When pressure is applied, the corpuscle is pressed flat. This flattening opens stretch-mediated $Na^+$ ion channels in the surface membrane of the neurone. As a result, $Na^+$ ions diffuse into the neurone and cause the generator potential that you can see in the graph

Figure 12.4 shows a cross-section through the layers of connective tissue and the neurone within a Pacinian corpuscle. It also represents the membrane potential of the neurone.

1   In its resting state (Figure 12.4a), a Pacinian corpuscle is round. In this condition, the stretch-mediated $Na^+$ channels are too small to let $Na^+$ ions pass through them and into the neurone. The neurone has a resting potential, represented on the graph.

2   When the layers of connective tissue and the neurone are compressed, the Pacinian corpuscle becomes flattened (Figure 12.4b). As a result, the surface membrane of the neurone is stretched. This causes its $Na^+$ ion channels to open and allow $Na^+$ ions to diffuse into the neurone. As a result, the membrane potential of the neurone changes. You can see this in the graph in Figure 12.4b, where a generator potential has been produced.

The generator potential causes an action potential. Once an action potential has been produced, a stimulated Pacinian corpuscle conducts an impulse along its length and stimulates a sensory neurone by releasing a neurotransmitter across a synapse.

**Q** **2** When the pressure on a Pacinian corpuscle is released, the neurone does not immediately become round again. Instead, it springs into an elongated shape at right angles to Figure 12.4b. Explain why this enables you to feel that the pressure has been released.

## The eye

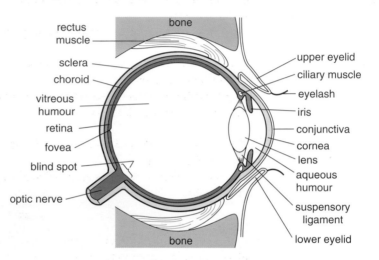

**Figure 12.5**
A cross-section through the human eye. The eye is a complex sensory organ, containing a variety of tissues. Only one of these, the retina, contains the receptors in which we are interested

Figure 12.5 shows that the human eye is a complex sensory organ, containing a variety of different tissues. Only one of these tissues, the **retina**, contains the receptors in which we are interested. Table 12.1 summarises the function of the different parts of the eye that can be seen in Figure 12.5.

Light is formed of particles, called photons. Each photon travels through transparent media in a light ray, which we represent as a straight line. When rays of light strike an object, they are reflected back at a predictable angle. When light rays pass from one medium to another medium of different density, the angle at which they travel is changed. This is called **refraction**. For example, light is refracted when it passes from air outside a building into the glass of a window and again when it passes from the glass in a window into the air inside a room.

**Q** **3** Name the surfaces within an eyeball at which light will be refracted.

There are millions of rays of light striking this page. They are reflected in countless directions and only some enter your eye as you read. Because these rays of light are from an object that is close to your eye, they are diverging (spreading out). If you look through a window at an object outside the room in which you are sitting, rays of light from distant objects enter your eye. Because these rays of light are from a distant

| Part of eye | Function |
| --- | --- |
| conjunctiva | ● protects the cornea against damage by friction (tears from the tear glands help this process by lubricating the surface of the conjunctiva) |
| sclera | ● protects eyeball against mechanical damage<br>● provides attachment surfaces for eye muscles |
| cornea | ● bends (refracts) light more than any other part of the eyeball<br>● allows passage of light |
| choroid | ● contains black pigment that absorbs light, thus preventing reflection of light within the eyeball<br>● contains blood vessels that supply other structures within the eyeball |
| ciliary body | ● has suspensory ligaments that hold the lens in place<br>● secretes aqueous humour<br>● contains ciliary muscles that enable the lens to change shape, during accommodation (focusing on near and distant objects) |
| iris | ● changes the size of the pupil, so controlling the amount of light passing to the inside of the eyeball<br>● contains pigment that absorbs light, reducing the amount of light passing to the inside of the eyeball |
| lens | ● bends (refracts) light<br>● by changing shape, focuses light from near objects (near accommodation) and distant objects (distant accommodation) on to the retina |
| retina | ● contains the light receptors – rod cells and cone cells<br>● contains relay neurones and sensory neurones that pass impulses along the optic nerve to part of the brain that controls vision |
| fovea (yellow spot) | ● a part of the retina that is directly opposite the pupil and contains only cone cells |
| blind spot | ● the part of the retina at which the sensory neurones that form the optic nerve leave the eyeball; it contains no light-sensitive receptors |
| aqueous humour | ● helps to maintain the shape of the anterior chamber of the eyeball |
| vitreous humour | ● supports the lens<br>● helps to maintain the shape of the posterior chamber of the eyeball |

**Table 12.1**
The functions of the major parts of the human eyeball shown in Figure 12.5

object, they are almost parallel. Figure 12.6 shows how we focus rays from near and distant objects by changing the shape of our lenses, in a process called **accommodation**. Contraction and relaxation of ciliary muscles changes the tension on the lens. When the lens is pulled thin, it bends light less. Our lenses are pulled thin when we focus the parallel rays of light from distant objects (**distant accommodation**). When the lens is allowed to revert to its fat shape, it bends light more. Our lenses are allowed to revert to their fat shape when we focus the diverging rays from near objects (**near accommodation**).

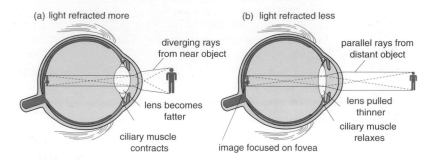

**Figure 12.6**
(a) During near accommodation, the ciliary muscles contract. This reduces the tension in the suspensory ligaments and the lens reverts to its fat shape. This bends light more, focusing the diverging light rays from a near object.
(b) During distant accommodation, the ciliary muscles relax. This increases the tension in the suspensory ligaments, pulling the lens much thinner. This bends light less, focusing the parallel rays from a distant object

If accommodation is successful, all the rays of light entering the eye from one part of an object will be focused at a single point of the retina. Unsuccessful accommodation results in blurred images on the retina. Some people produce clear images from distant objects but blurred images from near objects – they are long-sighted. Other people produce blurred images from distant objects but clear images from near objects – they are short-sighted. Both types of defect can be corrected with spectacles or contact lenses.

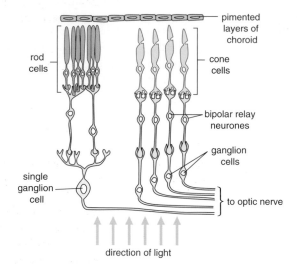

**Figure 12.7**
A cross-section of the human retina. Notice that light must pass through neurones before it strikes the retina

The human retina contains millions of light-sensitive receptor cells and the neurones with which they synapse. Figure 12.7 represents part of the human retina. Pigmented cells of the choroid are at the top of Figure 12.7. Although not strictly part of the retina, they have been drawn to remind you of their importance in absorbing light. The light receptors are shown in the layer of the retina below the pigmented cells. In this diagram, two types of receptor have been drawn. These are **rod cells** and **cone cells**. A layer of **bipolar relay neurones** lies beneath the layer of rod cells and cone cells and, finally, there is the lowest layer of sensory neurones (called **ganglion cells**). The axons of these ganglion cells make up the **optic nerve**, which passes impulses from the retina to the brain.

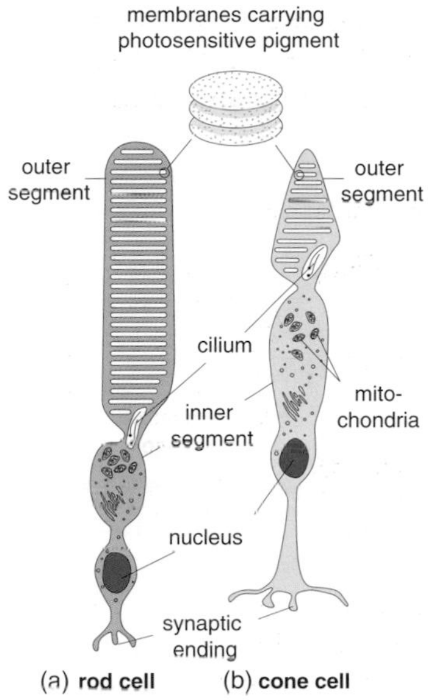

We need to look at the rod cells and cone cells first. Figure 12.8 shows
their structure in more detail and Table 12.2 summarises some important
differences between them.

**Figure 12.8**
(a) Rod cells and (b) cone cells
have an outer segment, with light-
sensitive pigment held in stacks of
membrane, an inner segment,
containing the cell's nucleus
together with organelles such as
mitochondria and ribosomes, and
a synaptic region

| Feature | Rod cells | Cone cells |
|---|---|---|
| approximate number in a human retina | • $120 \times 10^6$ | • $6 \times 10^6$ |
| distribution | • evenly throughout the retina<br>• absent from the fovea<br>• the only type of light receptor at the periphery of the retina<br>• absent from the fovea | • present mainly in the fovea (Figure 12.5) at a density of about $5 \times 10^4$ mm$^{-2}$ |
| shape of outer segment | • rod-shaped | • cone-shaped |
| sensitivity to light | • very sensitive to light, therefore operate even in dim intensities<br>• insensitive to colour (monochromatic vision) | • sensitive only to bright light, therefore operate only in bright light intensities<br>• sensitive to red light or green light or blue light |
| visual acuity | • produce poorly resolved images | • produce well-resolved images |
| light-sensitive pigments | • single pigment, called rhodopsin, in every rod cell<br>• stimulation of different combinations of the three types of cone cell produces a perception of colour (trichromatic vision) | • one of three types of iodopsin pigment in any one cone cell<br>• each type of iodopsin is sensitive to red light or to green light or to blue light |
| synapse with relay cell | • groups of rod cells synapse with one relay cell (retinal convergence) | • each cone cell synapses individually with a single relay cell |

**Table 12.2**
A comparison of rod cells and cone cells

Remember that receptors, such as rod cells and cone cells, convert a stimulus to a nerve impulse (transduction). Although rod cells and cone cells do this in a similar way, we only need to understand how transduction occurs in rod cells. In total darkness, the membrane of a rod cell is polarised, i.e. there is a potential difference between its cytoplasm and the surrounding fluid. As with neurones, this potential difference is caused by an imbalance in the concentration of $Na^+$ ions and $K^+$ ions between the cytoplasm and the surrounding fluid. In the dark, each rhodopsin molecule is made of a protein, called **opsin**, combined with *cis*-retinal. When *cis*-retinal absorbs light, it changes to a different isomer, called ***trans*-retinal**. In this form, the retinal cannot bind to opsin, so that the molecule of rhodopsin breaks down into *trans*-retinal and opsin. As a result, the surface membrane of the rod cell changes its permeability to $Na^+$ ions and $K^+$ ions, increasing the negativity of the cell's cytoplasm. It is this **hyperpolarisation** that causes a generator potential to be produced in the rod cell. As a result, the rod cell releases less neurotransmitter onto the relay cell with which it synapses.

$$Rhodopsin \xrightarrow{light} opsin + trans\text{-}retinal$$

**Q 4 Does the neurotransmitter released by rods have an excitatory or inhibitory effect on relay cells?**

If reduction in the release of neurotransmitter from a rod cell reaches a threshold, an impulse passes through the relay cell, across a synapse and down the ganglion cell. The likelihood of reaching a threshold in the relay cell is increased because several rod cells synapse with each individual relay cell. This is called **retinal convergence**. Although it increases sensitivity to dim light, retinal convergence decreases **visual acuity**. Look at Figure 12.9, which is a simple representation of the synaptic connections in the retina. Two cone cells are shown. Each of these cone cells synapses with a different relay cell that, in turn, synapses with a different ganglion cell. Stimulation of these two cone cells would result in two impulses passing along the optic nerve to the brain. Four rod cells are shown in Figure 12.9. All four rod cells synapse with the same relay cell, which synapses with one ganglion cell. Stimulation of any combination of these four rod cells would result in a single impulse passing along the optic nerve to the brain.

**Q 5 Suggest why you see a star in the night sky more clearly if you look to one side of it than if you look directly at it.**

Rhodopsin has to be resynthesised. Rod cells do this by combining retinal and opsin, using ATP produced by the mitochondria in their inner segments. This reaction occurs slowly compared to the breakdown of rhodopsin by light. In bright light, your eyes are light-adapted, meaning that most of your rhodopsin has been broken down into opsin and *trans*-retinal. If you go from bright light into dimly lit conditions, your vision will be poor until you have resynthesised enough rhodopsin for your rods to start working again. Your eyes have then become **dark-adapted**.

$$trans\text{-}retinal + opsin \longrightarrow rhodopsin$$
$$ATP \qquad ADP + P_i$$

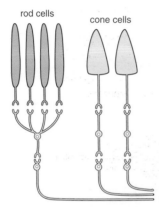

**Figure 12.9**
Simplified diagram to show retinal convergence in the retina. Cone cells synapse individually with relay and sensory neurones, increasing visual acuity (the ability of the brain to resolve images). Several rod cells synapse to the same relay cell. As a result, they collectively cause a generator potential in the relay cell in dim light conditions (summation). However, this pattern of synapses results in poor visual acuity

**Q** **6** Retinal is a derivative of vitamin A. Suggest the effect that a deficiency of vitamin A in the diet would have on vision.

## Skeletal muscles as effectors

Motor neurones stimulate glands and muscles into action. Because they produce a response to a stimulus, glands and muscles are known as **effectors**. In this part of the chapter, we will look at the structure of one type of muscle tissue, skeletal muscle, and the way in which it works.

### Structure of skeletal muscle

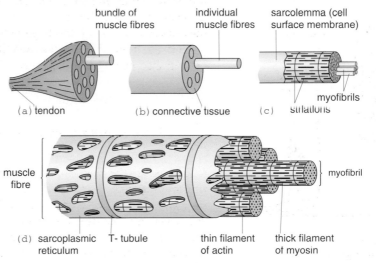

**Figure 12.10**
A skeletal muscle (a) is joined to the membrane covering a bone by inelastic tendons. Each muscle is made of hundreds of bundles of muscle fibres (b). Each muscle fibre (c) contains myofibrils (d)

Skeletal muscle is attached to the membranes that cover the bones of our skeletons by strong, inelastic **tendons**. When a muscle contracts, it pulls on the tendons and causes the movement of one of the bones to which the tendons are attached. Figure 12.10(a) shows an individual muscle cut across its length. It contains bundles of muscle fibres surrounded by connective tissue (Figure 12.10(b)). Each **muscle fibre** can be several centimetres long and is a single cell containing many nuclei. In Figure 12.10(d), you can see that its surface membrane is called the **sarcolemma**, its cytoplasm is called **sarcoplasm** and that it contains a network of membranes, called the **sarcoplasmic reticulum**. Along its length, the sarcolemma forms deep tubes into the sarcoplasm, called **T tubules**. As with any cell, the sarcoplasm contains a water-based suspension of proteins and dissolved inorganic ions surrounding a variety of cell organelles. However, most of the sarcoplasm is filled by parallel **myofibrils** (Figure 12.10(d)).

Figure 12.11 is an electronmicrograph of part of a single myofibril. It shows alternating bands across its length, called striations. These striations are caused by long molecules of protein, called filaments, which almost fill the sarcoplasm. There are two types of filament, one contains a protein called **actin** and the other contains a protein called **myosin**. Actin filaments are much thinner than myosin filaments, so they produce lighter striations. The actin filaments are attached to plates that cross the myofibril, called **Z lines**. The distance between two adjacent Z lines is called a **sarcomere**.

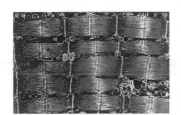

**Figure 12.11**
False colour transmission electronmicrograph of part of a myofibril, showing the characteristic striations

**Figure 12.12**
The striations in skeletal muscle are caused by filaments of two proteins: actin and myosin

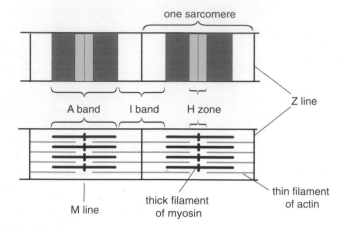

Look at Figure 12.12, which represents two adjacent sarcomeres in a myofibril. You can see how the actin filaments are attached to Z lines and extend into the sarcomeres on either side. Here, they overlap with the thicker myosin filaments, which are not attached to the Z lines. Because the myosin filaments are thicker than the actin filaments, they cause a darker striation. The striation caused by overlapping actin and myosin is called the **A band** and the striation caused by actin alone is called the **I band**. The part of the A band that contains only myosin, with no overlapping actin, is the **H zone**. Each myosin filament has a central thickening, which gives rise to the **M line**. Compare Figure 12.12 with Figure 12.11 and see if you can distinguish individual sarcomeres with their A bands, I bands, H zones and M lines in the electronmicrograph.

**Figure 12.13**
Cross-sections of a sarcomere to show the arrangement of myofibrils at three different places

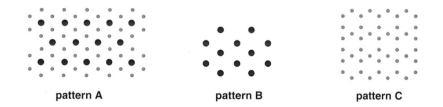

pattern A      pattern B      pattern C

**Q**   7   Figure 12.13 shows the arrangement of myofibrils in cross-sections through a sarcomere at three different places.
(a) What do these diagrams show about the arrangement of actin filaments?
(b) Use your knowledge of the distribution of actin and myosin to identify which part of the sarcomere is represented by each diagram.

**Extension box 1**      **Neuromuscular junction**

Skeletal muscle fibres are stimulated by individual motor neurones. The synapse between a motor neurone and its muscle fibre is called a **neuromuscular junction**. In Figure 12.14, you can see the end of the axon of a myelinated motor neurone. Like synaptic knobs (Figure 11.8), the swollen end of the motor neurone contains vesicles of neurotransmitter. In this case the neurotransmitter is always acetylcholine.

**Figure 12.14**
A neuromuscular junction: the synapse between a motor neurone and its target muscle fibre

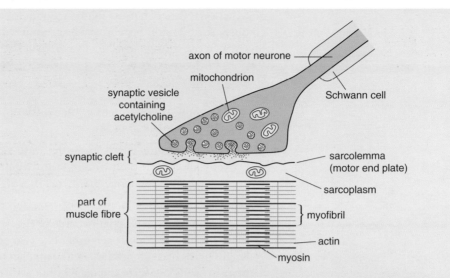

When an impulse reaches a neuromuscular junction, an influx of calcium ions causes synaptic vesicles to fuse with the presynaptic membrane at the end of the motor neurone and release their acetylcholine. The acetylcholine diffuses across the synaptic cleft and attaches to acetylcholine receptors on the postsynaptic membrane of the muscle fibre (called a **motor end plate**). In a similar way to nerve impulse transmission across a synapse, this causes depolarisation of the motor end plate. This causes a wave of depolarisation over the sarcolemma and down the T tubules of the sarcoplasmic reticulum. This wave of depolarisation changes the permeability of the sarcoplasmic reticulum to calcium ions and calcium ions move into the sarcoplasm.

## Sliding filament theory

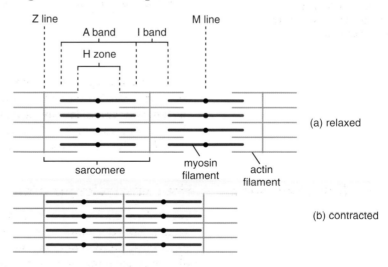

**Figure 12.15 (a), (b)**
During contraction, the length of individual sarcomeres gets less as actin filaments overlap more with myosin filaments. Notice that the length of the actin and myosin filaments stays constant.

Figure 12.15 shows what happens when a muscle fibre contracts. In (a) you can see the actin and myosin filaments in the adjacent sarcomeres of a relaxed muscle fibre. In (b) you can see the same muscle fibre during contraction. Notice that the filaments of actin and myosin have not changed their length. Instead, each sarcomere has been shortened by an increase in the overlap of actin and myosin filaments. In other words, during contraction the actin and myosin filaments slide over each other so that their overlap increases.

**Q** 8 **Explain what happens to the width of the A band and I band during muscle contraction.**

We now need to know what causes the actin and myosin filaments to slide over each other. To understand this, we first need to explore the structure of actin and myosin filaments in a little more detail. Figure 12.16 helps us to do this.

- Each actin filament is actually made of two strands of actin molecules that are twisted around each other like two pearl necklaces. Along the length of the actin molecules are the **myosin binding sites** at which myosin heads can form chemical bonds with the actin. Associated with these two strands of actin are two other proteins: globular **troponin** and two strands of **tropomysin**. Together, the troponin–tropomysin–actin complex conceals the binding sites of the actin molecules.

- Each myosin filament is made of bundles of myosin molecules. Each myosin molecule has a globular 'head' and a rod-like 'tail'. In a myosin filament, myosin molecules are bundled together so that the 'tails' form a central stalk from which the globular 'heads' stick out to the side.

**Figure 12.16**
In a single myosin filament, the myosin 'tails' form a central stalk and the 'heads' stick out to the side. These 'heads' attach to specific sites on the actin filament. The two actin molecules in each actin filament give the appearance of a twisted pearl necklace

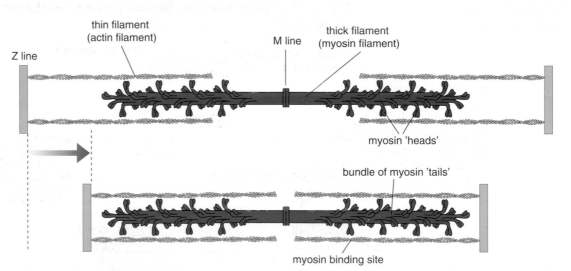

When muscle fibres are stimulated by the nervous system, the myosin 'heads' attach to their binding sites on the actin filaments to form temporary cross bridges. During one contraction, each cross bridge attaches and detaches up to 100 times. Following detachment, a new cross bridge is formed slightly further along the actin filament. This action is summarised in Figure 12.17 and is possible because:

1 myosin 'heads' contain an ATPase enzyme that hydrolyses ATP into ADP and inorganic phosphate (ATP → ADP + P$_i$)

2  calcium ions, which enter the sarcoplasm during the wave of depolarisation, bind to the troponin–tropomysin–actin complex and cause the myosin binding sites to be revealed and they stimulate the ATPase activity of the myosin 'heads', so releasing energy

3  using energy released from the hydrolysis of ATP, each myosin 'head' forms a new cross bridge with its exposed binding site on an actin filament

4  using energy released from the same hydrolysis of ATP, each myosin 'head' then undergoes a change in shape. It bends its 'head', causing the myosin to pull on the actin filament (the 'power stroke'). As a result, the Z line moves closer to the myosin filament. During this process, ADP and $P_i$ are released from the myosin

5  as a new molecule of ATP attaches to the myosin 'head', the actin–myosin cross bridge breaks

6  energy released by hydrolysis of the newly attached molecule of ATP changes the shape of the myosin 'head' to its original position. It is now ready for the next attachment to the actin filament and its next power stroke.

So long as enough calcium ions and ATP are present in the cell, the cyclic process of ATP binding to myosin, ATP hydrolysis, attachment of myosin heads to receptor sites on actin, change of angle of myosin head and detachment of myosin heads from receptor sites on actin can continue. In this way, the myosin 'heads' act like tiny oars that pull the actin filaments into the A band.

**Figure 12.17**
The cycle of events during contraction of a myofibril

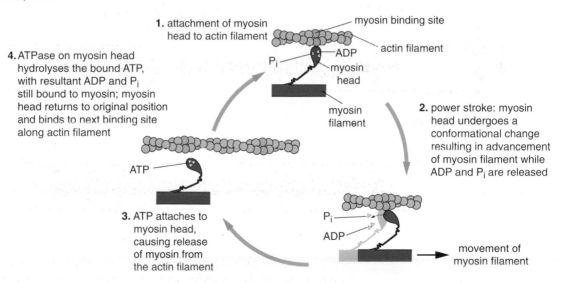

When stimulation by the nervous system stops, calcium ions are taken up by the sarcoplasmic reticulum. As a result, cross bridges will not reform after the myosin head has detached from its actin. The muscle will relax. If the supply of ATP runs out, attached cross bridges will not be able to detach. The muscles will become stiff and unable to relax. An extreme example of this is called rigor mortis, which occurs after death when supplies of ATP in the muscles run out.

**Q** 9 Give the role of (a) calcium ions and (b) ATP during the contraction of a myofibril.

During vigorous muscle contraction, stored ATP is used within 10 seconds. Since ATP is needed for muscle contraction, this depletion of ATP stores could stop further contraction. This does not happen because a compound called **phosphocreatine** is used to generate more ATP. Phosphocreatine is stored in muscle cells. It breaks down in a stepwise manner to creatine and an inorganic phosphate group, releasing energy in the process. This reaction is coupled to the resynthesis of ATP from ADP.

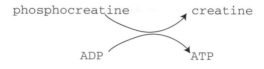

## Extension box 2

### Energy in active muscle cells

Active muscle cells use ATP at a rapid rate. During exercise, they resynthesise ATP using one of several alternative pathways. We have seen one of these pathways, in which the breakdown of a molecule of phosphocreatine provides a phosphate group and the energy to attach it to a molecule of ADP. This enables muscle cells to continue exercise until other slower pathways begin to form ATP.

The first of these pathways is the breakdown of glycogen that is stored in muscle cells. This provides glucose, which is broken down by aerobic respiration to produce ATP (see Chapter 5). After that, glucose and fatty acids from the bloodstream provide the substrate for aerobic respiration, with fatty acids being the more long-lasting substrate.

During prolonged muscle contractions, the gas-exchange and circulatory systems often fail to deliver oxygen at a rate that will support aerobic respiration. If this happens, anaerobic respiration will occur. As you learned in Chapter 5, anaerobic respiration forms lactate. This causes the cramps experienced during prolonged exercise, especially in untrained people.

## Slow and fast muscle

Often we think of muscle contraction occurring only when we perform rapid, energetic actions, such as those involved in exercise. However, muscles that retain our posture contract continuously, often without our being aware of them.

**Fast muscles** have a relatively poor blood capillary supply and their muscle cells lack myoglobin and have relatively few mitochondria. These fast muscle cells contract powerfully and quickly but can only do so for short periods of time. In contrast, **slow muscles** have a relatively rich blood capillary supply. Their muscle cells appear bright red because their sarcoplasm is packed with myoglobin and they have large numbers of mitochondria. These characteristics of slow muscle cells enable them to generate ATP by aerobic respiration. Thus, although they can only contract relatively slowly, they are able to do so for long periods of time. Table 12.3 summarises these, and other, differences between fast and slow muscle.

**Q 10** Suggest why the breast meat of cooked turkey is light in colour whilst the leg meat is dark.

| Feature | Fast muscle | Slow muscle |
|---|---|---|
| Functional | | |
| ● role in body | rapid, powerful movements | postural endurance |
| Structural | | |
| ● diameter of fibres | large | small |
| ● number of capillaries | few | many |
| ● amount of sarcoplasmic reticulum | high | low |
| ● number of mitochondria | fewer | more |
| Mechanical | | |
| ● speed of contraction | fast | slow |
| ● rate of pumping $Ca^{2+}$ ions | high | slower |
| Biochemical | | |
| ● ATPase activity | high, so split ATP quickly | low, so split ATP slowly |
| ● source of ATP | glycolysis (anaerobic) | electron transport (aerobic) |
| ● glycogen content | high | low |
| ● myoglobin content | low | high |
| ● rate of fatigue | fast | slow |
| Location | arms and legs (running and throwing) | back and neck (postural muscles) |

**Table 12.3**
A comparison of fast and slow skeletal muscle

## Summary

- Receptors convert specific stimuli into nerve impulses. They do so because a stimulus changes their resting potential, causing a generator potential in the receptor cells. Receptors pass their impulses across synapses to relay neurones or to sensory neurones.

- Pacinian corpuscles are sensitive to changes in pressure. Rod cells and cone cells are receptors in the retina of the eye that are sensitive to light.

- The eyeball is adapted to refract (bend) light so that it is focused onto light-sensitive cells in the retina. Refraction occurs at the junction of the conjunctiva and cornea, cornea and aqueous humour, aqueous humour and lens, lens and vitreous humour. By changing the shape of the lens, light from distant and near objects can be refracted differently so that it is focused onto the retina. This is called accommodation.

- There are two types of light-sensitive cells in the retina. Cone cells are relatively insensitive to light but can detect one of three colours, producing trichromatic vision. Rod cells are more sensitive to light but insensitive to colour, producing monochromatic vision.

- Rod cells produce a generator potential when light bleaches a pigment, called rhodopsin, which is present in their intracellular membranes. Since several rod cells synapse with an individual relay cell, they collectively cause depolarisation of the relay cell. This process, called summation, increases the sensitivity of the retina to light but reduces the visual acuity produced by rod cells.

- Cone cells synapse individually with individual relay cells, resulting in high visual acuity.

- Skeletal muscle cells have striations across their length, resulting from the distribution of filaments of two types of protein: actin and myosin. When stimulated at or above a threshold level, calcium ions flood into muscle cells causing actin and myosin filaments to slide over each other, increasing their region of overlap. As a result, the entire muscle cell contracts.

- The hydrolysis of ATP provides the energy for contraction. The breakdown of phosphocreatine, which is stored in muscles, is used to regenerate ATP.

- There are two types of skeletal muscle fibres. Fast muscle fibres are adapted to contract powerfully and quickly but can only do so for short periods of time: they are common in the legs and arms. In contrast, slow muscle fibres are adapted to prolonged contraction without fatigue: they are common in postural muscles, such as those in the back and neck.

# Assignment

## Going for Gold

Lynford Christie (Figure 12.18) was a world-class sprinter and a great athlete. Paula Radcliffe is not a sprinter, but she has the endurance to keep going over 10 000 metres. What is it that made Lynford a sprinter and makes Paula a distance runner? Part of the answer clearly lies in training, but part relies on the physiology of their muscles. In this assignment, we will look at muscle in a little more detail, and see just what it is that allows different runners to excel over different distances. The topic of muscle contraction brings together material from different parts of the specification. Before you start work on this assignment, it would be a good idea to look up the following key topics, either in your notes or in your textbook:

- respiration

- haemoglobin and the transport of oxygen

- nerve impulses

**Figure 12.18**
Sprinters, such as Lynford Christie, have a higher proportion of fast fibres in the muscles of their arms and legs than distance runners, such as Paula Radcliffe. In a sprinter, about 60% of the muscle fibres in some muscles are fast fibres. In an endurance athlete, around 80% of the fibres in the same muscles are slow fibres

Table 12.4 shows some of the differences between fast and slow muscle fibres and the nerves that supply them.

|  | Fast muscle fibres | Slow muscle fibres |
| --- | --- | --- |
| Features of muscle fibres | ● Few mitochondria<br>● Fibres white in colour because they contain very little myoglobin and none of cytochrome pigments involved in the electron carrier system<br>● Contain many glycogen granules | ● Many mitochondria<br>● Fibres red in colour because they contain myoglobin and cytochrome pigments involved in the electron carrier system<br>● Contain few glycogen granules |
| Nerve supply to muscle fibres | ● Large nerve fibres, approximately 15 μm in diameter | ● Small nerve fibres, approximately 5 μm in diameter |

Table 12.4

In this chapter, you have seen that energy in the form of ATP is made available to contracting muscles in several ways. These are:

● From that already formed and stored in the muscle cells. This is only a small amount, enough to allow muscle contractions to take place for about 3 seconds. ATP does not run out immediately after this, however, because it is reformed from the breakdown of phosphocreatine. In a trained athlete, the **ATP/phosphocreatine** system can supply enough energy to last for about 10 seconds of vigorous muscle contractions.

● From **glycolysis**. You have already seen that glycolysis is not very efficient at producing ATP. It does have a big advantage over aerobic respiration, however. It can produce ATP rapidly. Glycolysis will allow contractions for up to about 60 seconds. After this, lactate produced as a waste product builds up to concentrations that cause cramp and fatigue.

● From **aerobic respiration**. Krebs cycle and the electron transport system produce large amounts of ATP, which allow exercise to continue for much longer periods of time.

However, these processes take rather longer, so cannot produce ATP instantly. They will only continue for as long as there is enough oxygen supplied to the muscle.

1  (a)  By which of these three ways is ATP produced in slow muscle
       fibres?

*(1 mark)*

   (b)  Explain the evidence from the table that supports your answer.

*(2 marks)*

2  Suggest an explanation for the difference in the number of glycogen
   granules in the two types of muscle fibre.

*(2 marks)*

3  Maximum effort in jumping and throwing events is required for a
   very short period of time. Explain the advantage to athletes in such
   events of having muscle with a large number of fast muscle fibres.

*(1 mark)*

4  Myoglobin is very similar to haemoglobin, although there is one big
   difference in its chemical structure. Myoglobin molecules are made
   up of a single subunit, not the four that we have in a molecule of
   haemoglobin. However, it still functions as a respiratory pigment,
   and is able to combine with and release oxygen. Look at
   Figure 12.19. It shows dissociation curves for human myoglobin and
   haemoglobin.

**Figure 12.19**
This graph shows
dissociation curves for
human myoglobin and
haemoglobin

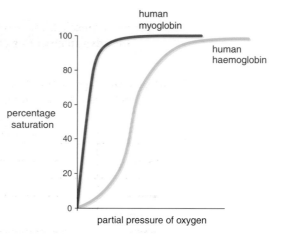

Use the graph to:

   (a)  explain why myoglobin is saturated with oxygen in resting slow
        muscle fibres;

*(3 marks)*

   (b)  explain how the presence of myoglobin would enable
        contractions in slow muscle fibres to take place for a long period
        of time.

*(2 marks)*

5  Explain how the nerve supply to fast muscle fibres could be an
   advantage to a competitor in an event such as fencing, where fast
   reactions are essential.

*(2 marks)*

# Examination questions

1  (a)  What is the role of phosphocreatine in providing energy for muscle contraction?

*(1 mark)*

   (b)  The table shows some differences between slow and fast muscle fibres.

| Slow muscle fibres | Fast muscle fibres |
|---|---|
| Enable sustained muscle contraction to take place | Allow immediate, rapid muscle contractions to take place |
| Many mitochondria present | Few mitochondria present |
| Depend mainly on aerobic respiration for the production of ATP | Depend mainly on glycolysis for the production of ATP |
| Small amounts of glycogen present | Large amounts of glycogen present |

   (i)  Explain the advantage of having large amounts of glycogen in fast muscle fibres.

*(2 marks)*

   (ii)  Slow muscle fibres have capillaries in close contact. Explain the advantage of this arrangement.

*(2 marks)*

2  The diagram represents a longitudinal section through part of a myofibril from a skeletal muscle.

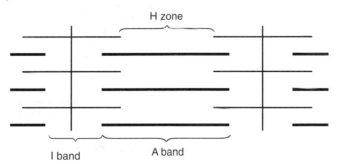

   (a)  The diagram shows the A band, and the H zone. Which one or more of these:

   (i)  contains actin but not myosin;

*(1 mark)*

   (ii)  shortens when the muscle contracts?

*(1 mark)*

   (b)  A slow muscle fibre is capable of sustained contraction such as is required for maintaining body posture. Suggest the advantage of the presence of large numbers of mitochondria in a slow muscle fibre.

*(3 marks)*

3 A person was instructed to close the left eye and stare with the right eye at a cross drawn on a plain white board. Different objects were then moved into the field of view from one side. The points where the objects were first seen and where the colours were first identified were recorded. The results are shown below.

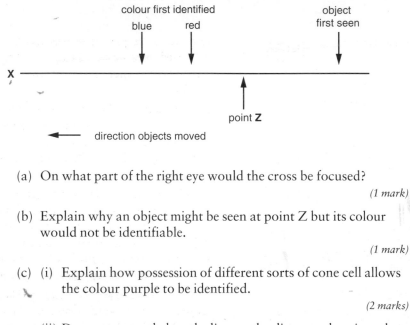

(a) On what part of the right eye would the cross be focused?

*(1 mark)*

(b) Explain why an object might be seen at point Z but its colour would not be identifiable.

*(1 mark)*

(c) (i) Explain how possession of different sorts of cone cell allows the colour purple to be identified.

*(2 marks)*

(ii) Draw an arrow below the line on the diagram showing where you would expect the colour purple to be first identified.

*(1 mark)*

# Homeostasis

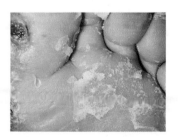

**Figure 13.1**
Humans live in and tolerate a wide range of physical conditions

**Figure 13.2**
Frostbite refers to tissue damage due to freezing, from the destructive effects of the formation of extracellular ice crystals. Frost-bitten parts need to be gently warmed

Humans are remarkably versatile in where they live on earth. They have extended their territories by use of clothing and by building shelters which protect them from unsuitable or extreme environmental conditions.

Regions inhabited by human populations have been determined by accessibility to land for food production and by the availability of water. Nevertheless, human populations show tolerance to wide temperature ranges, from a hot 50 °C in July in the Sahara desert to a cold −65 °C in January in Russia.

Despite this range of environmental temperature, the internal temperature of the human body is maintained within narrow limits. Specialised organs and systems of your body make continuous adjustments to ensure that all its cells are kept not only under optimum temperature conditions but have optimum water content, glucose concentration and oxygen concentration too.

The thermoneutral zone refers to the range of external (environmental) temperatures in which a naked 70 kg man can survive without needing to produce or lose heat. The thermoneutral zone has been defined as between 27 and 31 °C. The lowest temperature (27 °C) is known as the critical temperature. In reality, people usually wear clothes, so this critical temperature would be much lower than 27 °C. Any changes in body temperature are compensated for, mainly by adjustment in blood flow e.g. vasoconstriction or vasodilation.

If you live in a cool climate and pay a visit to a much hotter place, at first you will feel uncomfortable, then you begin to adjust, or acclimatise. However, when the body can no longer cope with excessively high temperatures, problems of heat stress arise. In some situations a person may feel dizzy or suffer from heat collapse. Recovery usually occurs quickly if the person is taken to a cool environment. Heat exhaustion is a more serious condition which may be due to dehydration. The most severe condition is known as heat stroke, characterised by a breakdown of the temperature regulation mechanisms altogether. If the core temperature of the body is 42 °C or higher, irreversible damage to the cells and proteins within those cells occur and the person is likely to go into a coma.

When the body can no longer cope with the effect of cold, problems of cold stress may result. This can cause severely reduced blood circulation which deprives the tissues of nutrients and their normal metabolic reactions fail. In a mild form, chilblains may develop. In more severe conditions, tissues actually freeze and this is known as frostbite.

## Why is homeostasis important?

The cells in any living organism will only function properly in the correct conditions. The rates of chemical reactions change according to environmental conditions and most organisms live in changeable environments. Temperature, for example, on a cold day in Britain might vary from 0 °C outside to 20 °C indoors. This variation could be a major problem if cells were at the mercy of changes in the external environment. The solution is to resist these changes, to keep conditions inside the cell constant, whatever may be going on outside.

Homeostasis involves mechanisms which keep conditions inside an organism within narrow limits and thus allow independence from fluctuating external conditions.

- Biochemical reactions are controlled by enzymes. Changes in pH and temperature affect the rate of enzyme-controlled reactions. Extreme temperatures and pHs can lead to denaturing of enzymes and other proteins.

- Water moves in and out of cells by osmosis. By maintaining a constant water potential in the fluid surrounding the cells, osmotic problems, which could lead to cellular disruption, are avoided.

## Keeping the internal environment constant

The blood cells are suspended in the blood plasma and other cells are surrounded by tissue fluid, which is derived from the plasma. A range of conditions in the blood (and therefore in the tissue fluid) is maintained close to the value at which cells function best. Keeping constant conditions in the tissue fluid around the cells is an example of homeostasis. The word homeostasis means steady state. Features of an organism's internal environment such as temperature, pH and the concentration of many dissolved substances have a set level, often referred to as the **norm**. Negative feedback is the process in which a fluctuation from this set level sets in motion changes, which return it to its original value. Although we talk about maintaining a constant internal environment, it is rarely constant, but fluctuates around the norm. Blood temperature in mammals is allowed to fluctuate only within very narrow limits, while blood glucose concentration can rise considerably above the norm without any harmful effects. Several different mechanisms are responsible for homeostasis but the basic features are always the same (Figure 13.3).

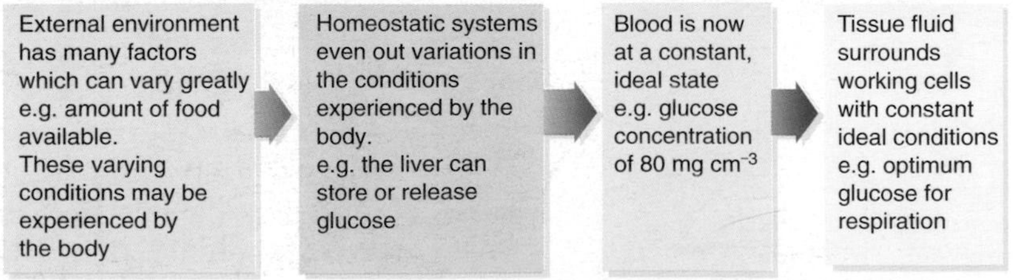

**Figure 13.3**
The basic features of homeostasis

| External environment has many factors which can vary greatly e.g. amount of food available. These varying conditions may be experienced by the body | Homeostatic systems even out variations in the conditions experienced by the body. e.g. the liver can store or release glucose | Blood is now at a constant, ideal state e.g. glucose concentration of 80 mg cm$^{-3}$ | Tissue fluid surrounds working cells with constant ideal conditions e.g. optimum glucose for respiration |

**Extension box 1**

# Regulation of blood cholesterol

## Cholesterol in the body

Most cholesterol found in the human body is made by liver cells. Some also enters the blood in foods such as butter and cheese. Despite concerns about health and high cholesterol concentrations, all cells need it as a component of membranes. The liver uses a lot to make bile salts, vital for fat digestion. Significant amounts are deposited in the skin, helping to make it waterproof, and some is used to make steroid hormones like oestrogen and testosterone. Acetylcholine, a neurotransmitter, is also made from it.

## Cholesterol in the blood

Cholesterol is insoluble in water. As it needs to pass to all cells it is carried in the blood in complexes called **lipoproteins**, mostly low density lipoprotein (LDL). A protein and phospholipid coating separates a ball of cholesterol from the aqueous environment (Figure 13.4).

## Why regulate cholesterol?

Cholesterol is needed by all cells, so must circulate in reasonable concentrations in the blood. However, at higher concentrations it tends to be deposited in the linings of artery walls. This narrows the artery and could make a heart attack or stroke more likely. Patients with an inherited inability to regulate blood cholesterol would not survive beyond childhood without treatment.

## Regulation

Liver cells are able to regulate cholesterol concentrations in their own cytoplasm as cholesterol inhibits one of the enzymes involved in its own synthesis. This is a typical example of negative feedback within the cell. However, by doing this, liver cells are also regulating the concentration in the blood. They export cholesterol to the blood, so when concentrations rise in the cells, concentrations in the blood rise as well. Just like other cells, liver cells also take in LDL, containing cholesterol, by endocytosis. This means that when blood cholesterol rises, cytoplasmic cholesterol increases too. In terms of cholesterol concentration, the blood can be seen as an enormous extension of the liver cell cytoplasm. By regulating concentrations inside the cell, they also regulate the amount in the blood. This is an unusually simple example of whole body homeostasis. One cell type both detects the change and brings about the response.

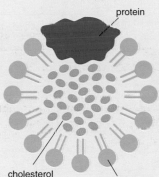

**Figure 13.4**
Lipoprotein

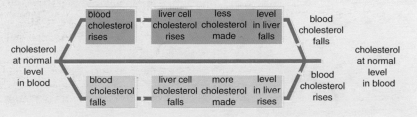

**Figure 13.5**
Summary of cholesterol regulation by the liver

## How is homeostasis brought about?

The work of a number of organs must be coordinated to achieve homeostasis. Information about the conditions in the body is continuously fed to the brain from sensory receptors around the body. For example, if the external temperature rises, temperature receptors in the skin send nerve impulses to the brain. In response, the brain initiates mechanisms that will lower the body temperature again. The same temperature receptors send nerve impulses to the brain when the temperature is back to normal. Homeostasis depends on the continual feedback of information (Figure 13.6).

**Figure 13.6**
In homeostasis, deviation from the norm acts as the signal that sets off the correction mechanism. This negative feedback keeps variable factors within the narrow ranges suitable for life

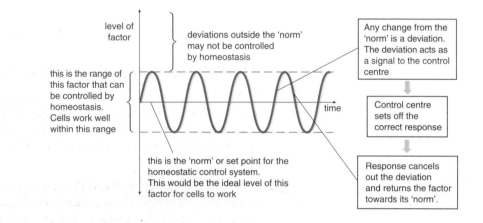

## Negative feedback

In biological systems, homeostasis is usually achieved by a process called negative feedback (Figure 13.7).

**Figure 13.7**
How negative feedback keeps a system in a stable condition

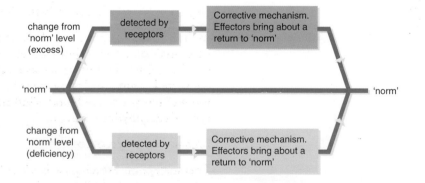

Negative feedback involves:

● a change in the level of an internal factor

● detection by receptors

● activation of effectors

● restoring the factor to its set point.

## What must be kept constant?

Some of the features of the blood and tissue fluid that must be kept within narrow limits are:

- the concentration of glucose

- temperature

- pH

- water potential

- the concentration of ions, such as sodium, potassium and calcium.

## The control of blood glucose concentration

### Blood glucose, supply and demand

In 1859 Claude Bernard, a French physiologist, was the first to recognise the importance of homeostasis in mammals. He studied variations in glucose concentrations in the blood of dogs and he found that the concentration remained remarkably stable despite dramatic variations in diet.

Humans eat a variety of foods that are sources of energy and raw materials. In normal circumstances, we obtain most of our energy by respiring glucose. Some vital organs, notably the brain, are unable to store carbohydrates and so cannot function for even a short period of time without glucose provided by the blood. Thus a lack of glucose in the blood can cause the brain to malfunction.

There are three sources of blood glucose:

- Digestion of carbohydrates in the diet

    In the UK, on average about 45% of our diet consists of carbohydrates. The carbohydrates we eat include sugars, starch and cellulose. Carbohydrates such as maltose, sucrose and starch are converted to glucose during digestion. Therefore the fluctuation in input of glucose to the blood is influenced by both the amount and type of carbohydrate we eat.

- Breakdown of **glycogen**

    This storage polysaccharide is made from excess glucose in a process called **glycogenesis**. Glycogen is particularly abundant in the cells of the liver and muscles. When needed, glycogen can be quickly broken down to release glucose.

- Conversion of non-carbohydrate compounds

    Glucose can also be produced from substances which are not carbohydrates. This process is called **gluconeogenesis**.

Glucose is oxidised during respiration and is broken down into carbon dioxide and water. During this complex series of chemical reactions, chemical potential energy in glucose is used to produce ATP. This in turn can be broken down to release energy for activities in the body. Some activities, for example vigorous exercise, demand a greater energy supply than others. This means that the rate of respiration varies, affecting how fast cells take up glucose from the blood.

The normal concentration of glucose in human blood is approximately 90 mg per 100 cm$^3$, and even after the largest carbohydrate meal it rarely exceeds 150 mg per 100 cm$^3$.

**Q** 1 **One person eats a bar of chocolate and another person eats the same mass of toast. Explain why there is a more rapid rise in blood glucose concentration in the person who has eaten the toast.**

## The mechanism of blood glucose control

### The pancreas

The **pancreas** plays a central role in the control of blood glucose concentration. The digestive functions of this organ are covered in Chapter 9, but here we are concerned with its endocrine role, the production of hormones.

The pancreas contains cells which are sensitive to blood glucose concentration and produces the hormones insulin or glucagon accordingly. These groups of cells are called **islets of Langerhans** (Figure 13.8). There are two types of islet cell. Glucagon is produced by α-cells and insulin is produced by β-**cells**. The islet cells are surrounded by other cells, which produce digestive enzymes.

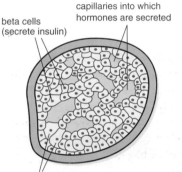

beta cells
(secrete insulin)

capillaries into which
hormones are secreted

alpha cells
(secrete glucagon)

**Figure 13.8**
Detail of an islet of Langerhans

### High blood glucose concentration

When blood glucose concentration becomes too high, β-cells in the islets of Langerhans detect this rise and respond by releasing insulin. This hormone travels to all parts of the body in the blood but mainly affects cells in muscles, liver and adipose tissue. Insulin has a number of effects on the body, all of which tend to lead to a reduction in the concentration of glucose in the blood.

- Insulin speeds up the rate at which glucose is taken into cells from the blood. Glucose normally enters cells by facilitated diffusion through protein carrier molecules in the plasma membrane. Cells have extra carrier molecules present in their cytoplasm. Insulin causes these carrier molecules to move to the membrane where they increase the rate of glucose uptake by the cell (Figure 13.9).

- It activates enzymes, which are responsible for the conversion of glucose to glycogen.

- It activates enzymes which promote fat synthesis.

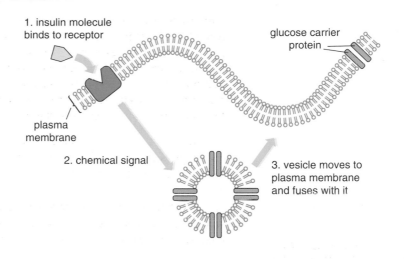

**Figure 13.9**
Insulin increases glucose uptake by increasing the number of carrier molecules in the plasma membrane

1. insulin molecule binds to receptor

glucose carrier protein

plasma membrane

2. chemical signal

3. vesicle moves to plasma membrane and fuses with it

## Low blood glucose concentration

If the concentration of blood glucose falls too low, this is detected by α-cells in the islets of Langerhans which then secrete glucagon. The main effect of this hormone on the body is to activate enzymes in the liver which are responsible for the conversion of glycogen to glucose. It also stimulates the formation of glucose from other substances such as amino acids. The glucose then passes out of the cells and into the blood, raising the blood glucose concentration. The control of blood glucose concentration is summarised in Figure 13.10.

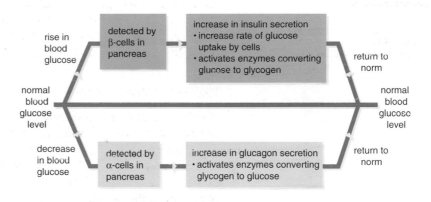

**Figure 13.10**
Summary of the homeostatic control of blood glucose concentration

**Q** 2 In the control of blood sugar concentration:
(a) what are the receptors
(b) how do the receptors activate the effectors
(c) what are the effectors?

## When homeostasis goes wrong – diabetes mellitus

Should the pancreas become diseased and fail to secrete insulin, or if target cells lose their responsiveness to insulin, the blood glucose concentration can reach dangerously high levels. The inability to control blood glucose concentration results in a condition called **diabetes mellitus**.

The blood concentration becomes so high that the kidney is unable to reabsorb back into the blood all the glucose filtered into its tubules. Diabetes mellitus is thus characterised by the excretion of large amounts of glucose in the urine. Other diagnostic features of this condition include a craving for sweet food and persistent thirst. The main diagnostic test is a glucose tolerance test in which the patient swallows a glucose solution and then the blood glucose concentration is measured at regular intervals. Graphs of the results of such a test from a diabetic and a person with normal glucose metabolism are distinctly different (Figure 13.11).

**Figure 13.11**
The results of a glucose tolerance test in a person with normal glucose metabolism and a diabetic

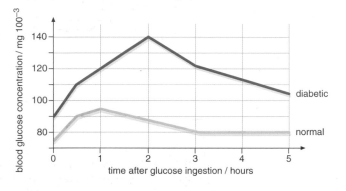

There are two main types of diabetes mellitus:

- **Type I** – also known as insulin-dependent or juvenile-onset. This usually occurs suddenly in childhood and results from the body being unable to make insulin. This is often caused by an **autoimmune reaction** (one in which the body's immune system attacks and destroys its own cells) which destroys the β-cells in the islets of Langerhans.

- **Type II** – also known as insulin-independent or late-onset. This usually occurs later in life and is more common than type I, accounting for more than 70% of the cases in the UK. It is often caused by a gradual loss in the responsiveness of cells to insulin.

### Treating type I diabetes

Before insulin treatment became available, about 75 years ago, type I diabetes would have been fatal within a year of diagnosis. Care has to be taken with the doses of insulin because an overdose causes too much glucose to be withdrawn from the blood, a condition known as **hypoglycaemia**. Therefore amounts of insulin given must match glucose intake and expenditure. Diabetics need to manage their diet and levels of exercise very carefully. All diabetics need to monitor their blood glucose concentration regularly. Easy-to-use biosensors are now available for this. Despite this most diabetics lead normal lives.

### Treating type II diabetes

Most type II diabetics control their blood glucose concentration by carefully regulating their diet, especially sugar intake and balancing this with the amount of exercise taken.

**Figure 13.12**
A person with type I diabetes has to inject with insulin regularly. Insulin cannot be taken orally because, being a protein, it would be digested in the alimentary canal

## Extension box 2

### Insulin patches

The chemical structure of insulin (Figure 13.13) shows two small polypeptide chains containing a total of 51 amino acids.

**Figure 13.13**
Chemical structure of insulin

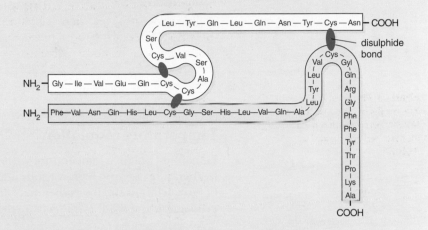

If insulin is swallowed, it is digested in the same way as other proteins in the diet. This means that the only way of using insulin at present is to inject it directly into the body. This is a straightforward procedure, but is not very pleasant.

Now there may be a better alternative for getting insulin into the body. Scientists have been experimenting with the use of ultrasound and skin patches. A skin area is first treated with ultrasound to disrupt the underlying fat tissue. This is necessary because insulin is not soluble in fat, so disrupting the fat tissue allows it to move in through the skin more easily. A patch containing insulin is then applied to the area, and can supply insulin for several days.

## Temperature control

Animal life can exist at almost all the temperatures encountered on the Earth's surface, from the extreme cold of the poles to the intense heat of the deserts. This is possible because it is the internal, not the external temperature that is important to an organism. If an animal's temperature falls too low biochemical reactions will be too slow for it to remain active. If it goes too high there is a risk of enzymes and other proteins being denatured. To a certain extent, all animals exert some control over their internal temperature.

### How heat is lost or gained

An organism's temperature changes as it gains or loses heat (Figure 13.14). If the heat gain is greater than the heat loss, the temperature of the organism will rise, and vice versa. To maintain a steady temperature, an organism must balance its heat gains and losses.

**Figure 13.14**
How an organism gains and
loses heat

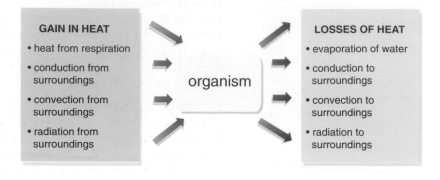

## Temperature regulation

Humans maintain a temperature within the range 36.1–37.8 °C. If the temperature rises too high, various internal mechanisms bring about loss of heat. If it falls too low, other mechanisms serve to increase heat production and reduce its loss from the body.

When we refer to body temperature, we actually mean **core body temperature**. In a human, the limbs and other extremities may be substantially cooler than 37 °C, but the core temperature usually remains constant (Figure 13.15).

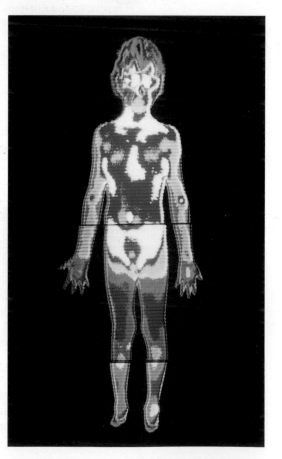

**Figure 13.15**
A thermal image of a person.
Areas of high temperature are
white, red or orange. Cooler
areas are green, blue or purple

### The role of behaviour

When we encounter a particularly warm or cold environmental temperature, temperature receptors in the skin send nerve impulses to the **voluntary** centre of the brain. So if we feel hot or cold we can decide to do something about it like changing position, changing clothing or turning the heat up or down.

### The role of physiology

If behaviour cannot deal with the problem, blood temperature starts to rise or fall. The **hypothalamus** of the brain detects the change in the temperature of the blood flowing through it and acts via the autonomic nervous system to initiate the appropriate response.

Temperature control in humans is achieved by negative feedback.

- When the core temperature rises too high, the resulting increase in blood temperature is detected by receptors in the hypothalamus. As a result the **heat loss centre**, which is also in the hypothalamus, sends nerve impulses to structures such as arterioles and sweat glands in the skin whose actions bring about the necessary fall in temperature.

- When the core temperature falls too low, this is also detected by receptors in the hypothalamus. This time, the **heat conservation centre** triggers mechanisms which conserve the body's heat or even generates more heat by actions such as shivering. The outcome, once again, is a return to the normal level (Figure 13.16).

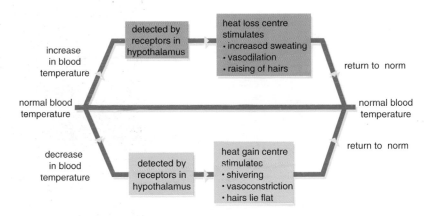

**Figure 13.16**
Summary of negative feedback in temperature control

### The role of temperature receptors in the skin

If the hypothalamus detects changes in temperature of the blood and initiates appropriate responses, what do the temperature receptors in the skin do?

While the hypothalamus detects temperature fluctuations inside the body, the skin receptors detect temperature changes at the surface. They enable you to feel whether the environmental temperature immediately outside your body is hot or cold. This information is sent by nerves to the voluntary centres of the brain and initiates voluntary activities such as jogging in severe cold or moving into the shade if conditions are hot.

## Physiological response to cold

When the hypothalamus detects a drop in blood temperature there are a number of responses that occur within the body.

- **Vasoconstriction** – This occurs when the arterioles leading to the capillaries in the outside layers of the skin narrow. This reduces the flow of blood through the capillaries (Figure 13.17). The capillaries do not contain muscle, so they cannot narrow. As less blood reaches the outside layer of the skin less heat is lost by radiation.

- **Hair raising** – This is when the hairs of the skin stand up. By raising the hair, it makes this hair layer thicker so that it traps more air and gives greater insulation. This has the greatest effect in hairy mammals. In humans where the hair is much less dense, although the hairs become upright, it has little effect. The most obvious feature is the presence of 'goose pimples'.

- **Shivering** – This is the rapid contraction and relaxation of muscles. When this occurs, the muscles give out much more heat than when the muscles are at rest.

- **Increased metabolic rate** – When we feel cold, the body secretes the hormone adrenaline. Adrenaline stimulates the cells to increase their metabolic rate. This results in an increase in heat production. Some mammals that live in cold climates for many months can increase their secretion of the hormone thyroxine. This increases the metabolic rate on a more permanent basis.

**cold conditions**

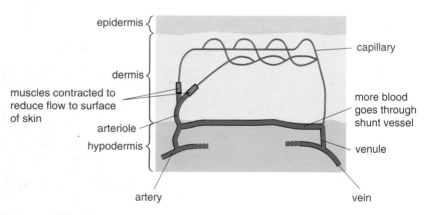

**Figure 13.17**
In cold conditions blood flows to the surface of the skin when the arterioles dilate during vasodilation

## Physiological response to heat

There are two main physiological responses that occur when the hypothalamus detects a rise in body temperature.

- **Vasodilation** – This occurs when arterioles leading to capillaries in the skin **dilate**. At the same time **shunt vessels,** which carry blood deep below the surface of the skin are closed off. This results in a greatly increased blood flow near the surface of the skin resulting in more heat being lost to the environment by radiation and conduction (Figure 13.18).

● **Sweating** – Sweat is a salty solution made by sweat glands. Evaporation of sweat from the skin's surface cools it. The evaporation of each gram of water requires 2.5 kJ of energy. Being furless, humans have sweat glands over the whole of the body, making cooling by this mechanism very efficient. In humans, sweat glands can lose up to 1000 cm³ hour⁻¹ of sweat. The efficiency of sweating depends on the humidity in the environment.

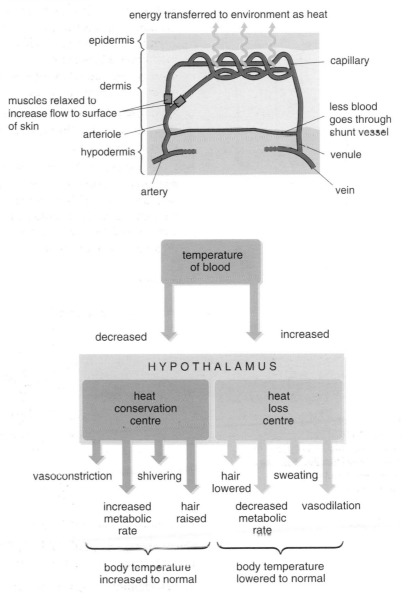

**Figure 13.18**
In warm conditions blood flow is now mainly in the shunt vessel away from the surface of the skin

**Figure 13.19**
Summary of body temperature control by the hypothalamus

### The structure and role of the skin in temperature regulation
In terms of area, the skin is the largest organ in the body. It is in direct contact with the external environment and so plays a central role in maximising and minimising heat loss (Figure 13.20).

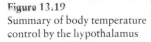

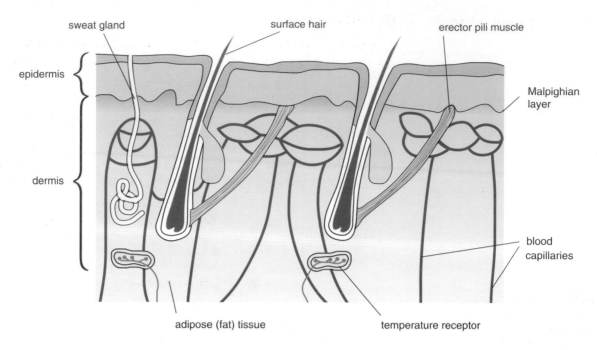

sweat gland

surface hair

erector pili muscle

epidermis

Malpighian
layer

dermis

blood
capillaries

adipose (fat) tissue

temperature receptor

**Figure 13.20**
The structure of human skin

Structurally the skin is divided into two layers, the outer **epidermis** and the inner **dermis**. Forming the boundary between the two is the **Malpighian layer,** the bottom-most layer of epidermal cells.

The cells of the Malpighian layer divide repeatedly by mitosis. As new cells are formed, the older ones get pushed outwards towards the surface, flattening as they do. After a time, their cytoplasm becomes full of granules and the cells die. Finally they become converted into scales of **keratin**. Keratin is a tough fibrous protein and gives the skin its protective properties and makes it waterproof. We are constantly losing epidermal cells. The dust in our houses is mainly made up of lost epidermal cells. The dermis is much thicker than the epidermis and contains structures such as nerve endings and blood vessels held together with connective tissue. Beneath the dermis is a region, which usually contains at least some **subcutaneous** fat. This fat storage tissue, called **adipose** tissue, provides vital insulation in humans.

Skin has several functions.

- It detects stimuli using cells that are sensitive to heat, cold, pressure and pain.

- It prevents excessive water loss or gain.

- It plays a role in temperature regulation by adjusting heat loss according to the conditions.

- It prevents entry of microorganisms.

**Extension box 3**

## Hypothermia

Hypothermia is the condition in which the body temperature falls dangerously below normal. It happens if heat energy is lost from the body more rapidly than it can be produced, for example when a person wearing inadequate clothing is subjected to prolonged cold.

As the body temperature falls, one of the first organs to be affected is the brain. This results in the person becoming clumsy and mentally sluggish. Since brain function is impaired, the victim may not realise that anything is wrong and so does nothing about it such as putting on more clothes. Indeed there have been examples when people even take off clothes.

As the body temperature falls, the metabolic rate falls too. That makes the body temperature fall even further, a case of positive feedback (Figure 13.21).

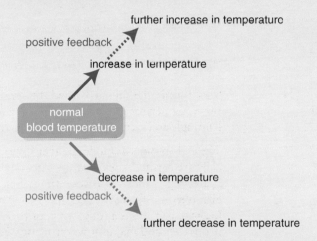

**Figure 13.21**
Positive feedback showing that after a deviation from the norm, the response is to take it further away from the norm

Death usually occurs when the body temperature drops to about 25 °C, though people have been known to survive lower body temperatures than this. The cause of death is usually ventricular fibrillation, a condition in which the normal beating of the heart is replaced by uncoordinated tremors.

Most at risk from hypothermia are babies and the elderly; babies because of their high surface area to volume ratio and undeveloped temperature regulation mechanisms, the elderly because their thermoregulatory mechanisms may have deteriorated through old age.

Deliberate hypothermia is sometimes used in surgical operations on the heart. The patient is cooled either by circulating the blood through a cooling machine or by placing ice packs in contact with the body.

By cooling the patient the metabolic rate is reduced and the demand for oxygen by the brain and other vital tissues is lowered. This allows the heart to be stopped without risk of the patient suffering brain damage through lack of oxygen. But the patient must not be cooled for too long or the tissues may be permanently damaged.

# Summary

- Homeostasis provides a constant internal environment and independence from fluctuating external conditions.

- 'Negative feedback' tends to restore systems to their original level.

- There are separate mechanisms that must be coordinated when there are departures from the original state in opposite directions.

- There are mechanisms that cause heat to be produced, conserved and lost.

- The hypothalamus monitors blood temperature and the autonomic nervous system coordinates temperature control in a mammal.

- Blood glucose concentration changes all the time depending on absorption of glucose from the digestive system and removal by active cells for respiration.

- Insulin and glucagon are hormones that control blood glucose concentration by activating enzymes involved in the interconversion of glucose and glycogen.

- Diabetes can be controlled using insulin or by manipulating carbohydrate intake.

# Assignment

## Writing synoptic essays

The idea of homeostasis is one that links many areas of an organism's biology and, as biologists, it is important that we understand links like this. One of the ways in which AQA Specification A assesses your ability to bring together different areas of the subject is with an essay. In this essay, marks are awarded not only for your scientific knowledge but also for selecting appropriate material from different parts of the specification. In order to write a good essay you need to practise this skill. In this assignment we will look at writing an essay. You will probably find it better to attempt it after you have completed this chapter and the next one. You will then have more knowledge of homeostasis.

Writing any essay, involves three steps. These are

- analysing the question so that you know exactly what you have to do

- planning your answer

- writing the essay

In this assignment we will concentrate on the first two of these steps – analysing and planning.

Here is your task.

Write an essay on the following topic:

The ways in which a mammal maintains constant conditions inside its body.

## Analysing the question

1 Start by looking at the instruction. In this case it requires you to "Write an essay" so you need to write out your answer in continuous prose. You shouldn't be using bullet points or writing in note form. If it helps to make a particular point, however, you can use a diagram, providing that it is relevant and adequately explained.

2 Look at the topic you are required to write about. The wording of the essay title tells you that you should confine your answer to mammals – so there is no place here for all that interesting stuff about how a crocodile maintains a constant body temperature. A crocodile is not a mammal. It also tells you that you have to look at maintaining constant internal conditions, so you should not, for example, include a lot of irrelevant detail about how the liver and the kidney function.

3 Finally, what about the mark allocation? The essay is worth 25 marks but, perhaps more important than this, you have about 40 minutes and that includes time for planning as well as writing.

## Planning your answer

Right. You know exactly what you have to do. The next step is to plan your answer. Planning is one of those awkward pieces of advice which everyone is given but few act upon. In an examination there never seems enough time for this. What we will try to do in this assignment is to show you how to produce an effective plan in a very short time.

You need a framework to help you to explore your knowledge of biology. Otherwise it is all too easy to concentrate on a few aspects of a particular topic. We can divide biology up in a number of different ways. Look at Table 13.1. It shows some of the ways in which we can do this:

| Prokaryotes | Protoctists | Fungi | Plants | Animals |
|---|---|---|---|---|
| Cell biology | Biochemistry | Physiology | Genes and genetics | Ecology |
| Gas exchange and transport | Nutrition | Homeostasis and excretion | Co-ordination and movement | Reproduction and growth |

Table 13.1

Each row in the table shows a different way of dividing up the subject. The first row is based on classification and shows the five kingdoms of living organisms; the second row shows the main areas of biology that make up this specification; the third concentrates on physiology. Now let's go back to our essay. We need to look at this table and choose the row that will provide us with the best framework for a plan. We can't really select the first row as the essay only concerns mammals. The second row is similarly of little use as we have to write about the ways in which a mammal maintains constant conditions inside its body. The third row seems the best option.

The next step is to use this as the basis of a brain storming exercise. Draw a table with the items in this third row forming your column headings. Now, try to write something under each heading – one way in which this topic is related to maintaining constant conditions. If you are stuck, don't worry. Move on to the next column. The object is to work as quickly as possible. Set yourself a limit of five minutes – you won't have more time than this in an examination. You might end up with something like Table 13.2.

| Gas exchange and transport | Nutrition | Homeostasis and excretion | Co-ordination and movement | Reproduction and growth |
|---|---|---|---|---|
| pH of the blood<br><br>Exercise and blood flow to the brain | | Temperature<br><br>Blood glucose concentration<br><br>Water balance | | Concentration of reproductive hormones such as FSH |

Table 13.2

We certainly haven't managed to write something in every column but if you look carefully at this table you will see that we have covered a good range of topics and we should be able to gain most of the marks available for demonstrating a range of knowledge.

## Writing the essay

You could try writing this essay but, before you do, look at the hints below:

- Structure your essay. Start with an introduction. In this case you could relate the title of the essay to homeostasis and, possibly, explain why maintaining constant internal conditions is so important. Follow this with the main content, devoting a separate paragraph to each of the different aspects you plan to write about.

- Make sure that you include enough detail. In practice this means that you can't afford to write more than about four lines without bringing in something that you have learnt as part of your A-level Biology course. Don't forget, you have only got about 40 minutes altogether.

- Keep it relevant. Don't get side-tracked and include things "just in case".

- Use the right language. To get your marks, you must use proper scientific terms.

# Examination questions

1 The diagram shows some important features of homeostatic mechanisms in the body.

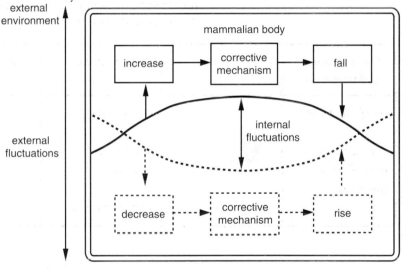

(a) Use the information in the diagram to help explain the importance of a human maintaining a constant internal temperature.

*(4 marks)*

(b) Describe the role of the hypothalamus in the regulation of body temperature.

*(3 marks)*

(c) Humans from different populations show different characteristics. People who have evolved in the far north of Canada tend to be shorter and stockier than those who evolved in tropical Africa.

(i) Explain the advantages of a short, stocky body form for a person living in very cold conditions.

*(4 marks)*

(ii) Explain how natural selection might account for these differences in body form.

*(4 marks)*

2 A person fasted overnight and then swallowed 75 g of glucose. The graph shows the resulting changes in the concentration of insulin and glucose in the blood.

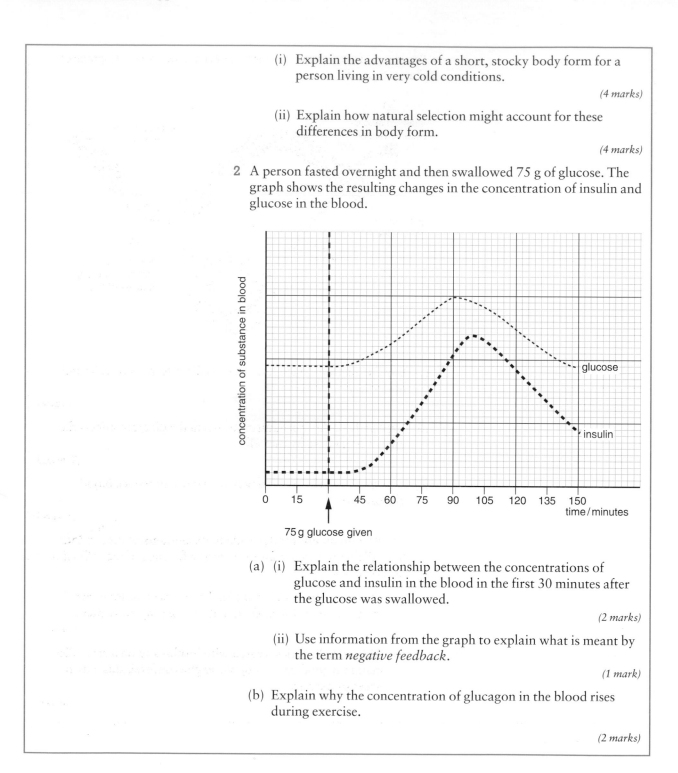

(a) (i) Explain the relationship between the concentrations of glucose and insulin in the blood in the first 30 minutes after the glucose was swallowed.

*(2 marks)*

(ii) Use information from the graph to explain what is meant by the term *negative feedback*.

*(1 mark)*

(b) Explain why the concentration of glucagon in the blood rises during exercise.

*(2 marks)*

The diagram shows how the hormone insulin increases the uptake of glucose by a cell.

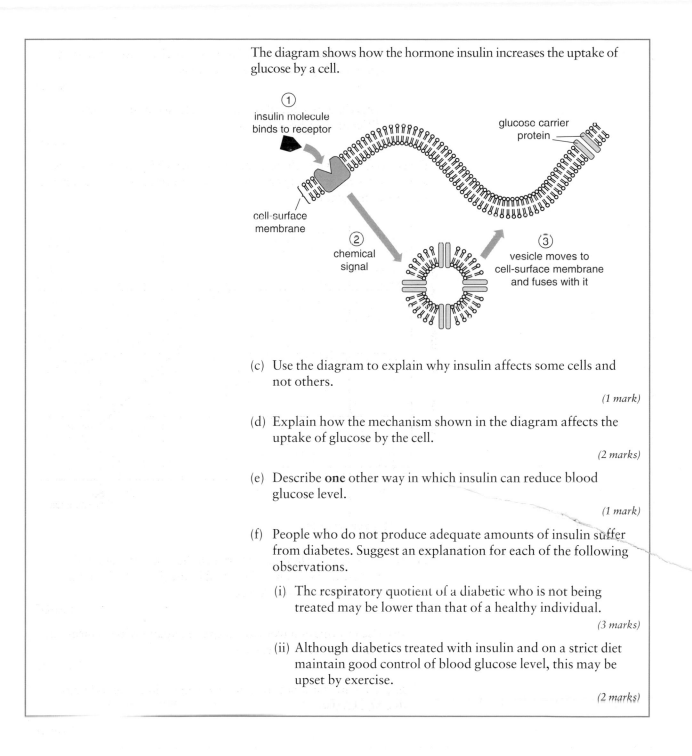

(c) Use the diagram to explain why insulin affects some cells and not others.

*(1 mark)*

(d) Explain how the mechanism shown in the diagram affects the uptake of glucose by the cell.

*(2 marks)*

(e) Describe **one** other way in which insulin can reduce blood glucose level.

*(1 mark)*

(f) People who do not produce adequate amounts of insulin suffer from diabetes. Suggest an explanation for each of the following observations.

(i) The respiratory quotient of a diabetic who is not being treated may be lower than that of a healthy individual.

*(3 marks)*

(ii) Although diabetics treated with insulin and on a strict diet maintain good control of blood glucose level, this may be upset by exercise.

*(2 marks)*

# Index

abiotic factors 74, 81, 84, 85, 88
absolute growth 166, 167
absorption 182–3, 189–92, 207
accommodation 249–50, 259
acetylcholine 235, 236, 239, 240,
    254–5, 267
acrosin 141
acrosome 133, 140–1, 154
actin 253–4, 255–7, 260
action potentials 230–3, 234, 240,
    241, 246, 248
active transport 146, 147, 154
    absorption 190, 191, 207
    neurones 228–9, 232
adenosine triphosphate (ATP) 94–5, 107
    muscle contraction 256–8, 260, 261
    retina 252
    sodium–potassium pump 232
    sperm 134
adolescence 163
adrenaline 276
ageing 163, 171–6, 177
alimentary canal 183–4, 189
'all or nothing' principle 231, 240
alleles 7, 11, 15–21, 26, 41–3, 218–19
allopatric speciation 57–8, 68
amino acids 39–40, 119, 171–2
    absorption 191
    diet 197, 207
    digestion 183, 187–8
    transamination 198
amylase 184
anaemia 205, 218
anaerobic respiration 101–2
anaphase 9, 10
Animalia 60, 66–7, 68
antagonists 235–6
anterior pituitary gland 150, 163,
    171, 175
antibodies 132, 146, 154, 191
athletes 207, 208
ATP see adenosine triphosphate
autoimmunity 176, 272
autonomic nervous system 235,
    238–40, 275, 280
autosomes 11
axons 227, 228, 233

basal metabolic rate (BMR) 173,
    177, 199–202, 208
bases, DNA 39
bias 33–4, 49
bile salts 188, 192, 267
biomass 105
biotic factors 74, 81, 85, 88
birth 150, 151, 154
blastocysts 142–3, 144, 154
blastula 67, 69
blood groups 17–18
blood pressure 240
blood volume 149, 154
BMR see basal metabolic rate
body fat 162, 168, 170
body mass 162, 164, 168, 173, 177,
    200–1

body temperature 265, 274–9
Bohr shift 216–17, 222
brain 237–8, 239, 240, 279
breasts 148, 153–4, 168, 169
breeding potential 55, 67
bronchi 238
brown fat 103–4

calcium ions 195, 199, 205–7, 208
    muscle contraction 257, 260
    synaptic transmission 234–5
carbohydrates
    diet 194–5, 207, 208
    digestion 183, 184–5
    homeostasis 269–70, 280
carbon cycle 115–19, 125
carbon dioxide 116, 118–19, 125
    haemoglobin 216–17, 221
    transport 220–2
carbonic anhydrase 220, 222
cardiac output 149, 173, 177
cardioacceleratory centre 239–40
cardiovascular system 148–9, 154
cervix 135, 136, 140, 150
chance 33–4, 49
chi-squared test 23–5, 26
chiasmata 9, 12–13, 26
childhood 163, 168
chloride ions 189–90, 221–2, 229
chloroplasts 95
cholesterol 267
chorophyll 95
chromatids 7, 8, 9, 12, 13, 26
chromosomes 7–8, 10–14, 21–3, 26,
    56
    mutations 176, 218
ciliary muscles 248–50
circulation, fetal 151–2
climax community 85, 88
codominance 16–17, 18
codons 39–40
coenzymes 100
collagen 171–2
communities 74
concentration gradients 147
conditioned reflexes 193, 194
conduction, nerve impulses 231–4
cone cells 246, 250–2, 259–60
continuous variation 32
contraception 132, 155–7, 204–5
contraction, muscle 255–8, 260
core body temperature 274–5
corpus luteum 138, 139, 143, 150–1,
    171
counter-current systems 147
cross-sectional studies 164–5, 177
crossing over 13, 37–8

data collection 33–4, 49
decomposers 104, 115
degrees of freedom 25
dendrites 227, 228
denitrifying bacteria 120
depolarisation 230–3, 235, 255, 260
development 163, 168–76, 177
diabetes mellitus 271–2, 280
diet 170, 173, 194–5, 206–7, 208
differentiation 144

diffusion 146, 147, 191, 207, 228–9,
    233–4
digestion 182, 183–8, 192–4, 207
dihybrid inheritance 19–21, 26
dipeptides 187, 188, 191
diploid cells 8, 9, 14, 26
directional selection 45
discontinuous variation 32
discrete data 32
disruptive selection 45
dissociation curves 215–17, 222–3
diversity 79–83, 88
DNA 7, 38–9, 176, 218
dominance 7, 15–16, 20–1, 26
dorsal root ganglion 236, 237
Down's syndrome 11
Drosophila melanogaster 15, 17, 24,
    31–2, 40–1, 42

ecological pyramids 104–5
ecology 73
ecosystems 74, 104–7
egestion 183
egg cells 8, 136–7
elastin 171–2
electron transport chain 101
embryos 142, 143, 144
emulsions 188, 192
endocrine system 193, 208
endocytosis 235, 267
endometrium 136, 139, 140, 142,
    145, 150
endopeptidases 187, 188
energy 93–107, 195–7, 199–202, 258
enterokinase 187
environments 74, 81, 85, 88
enzymes 183, 184–8, 192, 266,
    270–1, 280
epithelium 146, 147, 185, 189–92
essential amino acids 197–8, 207
Euglena 62–3
eukaryotes 60, 61, 68, 69
exercise 162, 173, 202, 206, 207
exopeptidases 187, 188
eye 238, 246, 248–52, 259

facilitated diffusion 146, 147, 190,
    191, 270
fallopian tubes 135, 136, 140, 142,
    143
family 59, 68
fast muscle 258–9, 260, 261
fat 267, 270
    see also brown fat; lipids
    basal metabolic rate 201
    obesity 206
    puberty 168, 170, 177
    skin 278
fatty acids 183, 188, 192
female reproductive system 135–40
fertilisation 132, 135, 136, 140–4
fertility 132, 174–5, 177
fetal haemoglobin 146, 147, 219–20,
    222
fetuses 144, 145–52
flagella 133, 134
follicle-stimulating hormone (FSH)
    138–40, 143, 150, 154, 155

menopause 175
puberty 169, 170–1
follicles 137–8, 139, 143
food 194–9
food chains/webs 104, 107
foramen ovale 152, 154
frequency distribution 31–3, 49
fruitfly see Drosophila melanogaster
FSH see follicle-stimulating hormone
fungi 60, 63–4, 68

gall bladder 184, 188, 194
gametes 133–4, 136–9, 154
gametophytes 65, 68–9
gaseous exchange 143, 146, 152
gastric juice 187, 188, 189, 192, 194, 208
generator potential 246, 247, 252, 260
genes 7, 14, 39, 176
genotype 13, 14, 26
genus 59, 68
gestation 144, 151
glucagon 270–1, 280
glucose
    absorption 190, 191
    collagen 171–2
    digestion 183, 184–5, 188
    exercise 207
    homeostasis 269–73, 280
glycerol 183, 188, 192
glycogen 207, 269, 270, 280
glycolysis 100
gonadotrophins 155, 170–1, 177
grana 95
growth 163–5, 166–7, 177, 203, 204
growth hormone 163, 169, 170–1, 177

habitats 75
haemoglobin 146, 147, 152, 204, 205, 214–22
haemophilia 21–3
haploid cells 8, 9, 14, 26
    gametogenesis 134, 136, 137
Hardy–Weinberg principle 41–3, 49
hCG see human chorionic gonadotrophin
heart 148–9, 152, 238
heart rate 149, 173, 214, 239–40
heat regulation 149–50, 265, 273–4, 275, 276–7, 279
height 162, 164–5, 166–7, 177
heterogametic sex (males) 14
heterotrophs 182
homeostasis 265–80
homogametic sex (females) 14
homologous chromosomes 8, 11–12, 26, 56
hormones
    ageing 176
    blood glucose 270–3, 280
    contraceptives 155–7
    digestion 193–4
    female reproductive system 138–40
    lactation 153–4
    menopause 175
    pregnancy 143, 146, 150–1, 153, 154
    puberty 168–71, 177
    temperature control 276

human chorionic gonadotrophin (hCG) 143, 146, 150–1
hybridisation 55, 57, 67
hydrochloric acid 187, 192
hydrogencarbonate ions 220–2
hydrolysis 183, 184, 187, 207
hyperpolarisation 230, 235, 252
hyphae 63, 64
hypothermia 279

implantation 142–3
impulses 226, 231–4, 240, 246, 259
independent assortment 12, 37
index of diversity 79–80
infants 163, 177
infertility 132, 143–4, 175
ingestion 183
inheritance 13–25, 26, 40, 49
inhibitory synapses 235
insulin 270–3, 280
intercostal muscles 238
interstitial cell stimulating hormone 169, 170
intestines 182, 183–5, 187–8, 189–92, 193–4, 207
ion channels 228–33, 234–5, 241, 247
ions, absorption 189–91
iris 238, 248, 249
iron 195, 199, 204–5, 206–7, 208
islets of Langerhans 270–1

Joules 195–7

keratin 278
kingdoms 59, 68
Klinefelter's syndrome 10
Krebs cycle 100–1

lactation 151, 153–4, 203, 206, 208
lacteals 192
lactose 185–7, 195
large intestine 183–4, 189
lens 248, 249–50, 259
LH see luteinising hormone
lichens 108–10
light-dependent reactions 95–7
link reaction 100
lipase 188
lipids 188, 192, 194–5, 208
    see also fat
lipoproteins 267
liver 183–4, 188, 198
locus 14
longitudinal studies 164–5, 177
lumen, alimentary canal 185, 189
lungs 152, 154, 216, 222
luteal phase 138, 139–40
luteinising hormone (LH) 139–40, 154, 155, 170–1

malaria 47–8, 49, 218–19
male reproductive system 132–5
Malpighian layer 278
maltose 184–5, 188
mark–release–recapture technique 78–9, 88
medulla, brain 239, 240
meiosis 9–13, 26, 37–8, 56

fertilisation 141
gametogenesis 134, 136, 137, 154
membrane, neurones 228–30
Mendelian inheritance 13–21
menopause 174–5, 177
menstrual cycle 139–40, 154, 168, 170–1, 174
menstruation 139–40, 150, 169, 178, 204–5
metabolic rate 199–202, 208, 276, 279
metabolism 173, 176, 194–5, 197–9
metaphase 9, 10
micelles 192
microvilli 146, 189, 192
milk 153–4, 185–7, 195, 203
minerals 194–5, 208
mitosis 9, 38
    fertilisation 141, 142
    gametogenesis 134, 136–7, 154
monoglycerides 183, 188, 192
monohybrid inheritance 15–18, 26
motor neurones 227–9, 235–7, 239, 240, 254–5
muscle 168, 177, 201, 253–9, 260, 261
mutations 38–40, 43–4, 46, 49, 218–19
mycelium 63, 64
myelinated neurones 174, 228, 233–4, 240, 254–5
myofibrils 253–4, 255, 257
myometrium 136
myosin 253–4, 255–7, 260
myxomatosis 45

natural selection 43–8
negative feedback 175, 266, 267, 268, 275, 280
nervous system
    ageing 174, 177
    autonomic 238–40
    digestion 193, 208
    reflexes 236–8
    temperature control 275, 280
neuromuscular junctions 242, 254–5
neurones 172, 174, 226–40, 247, 254–5
neurotransmitters 226, 234–6, 239–40, 248, 254–5, 267
nitrogen cycle 119–24, 125
nodes of Ranvier 227, 228, 233, 240
noradrenaline 235, 239, 240
normal distribution 34–6, 49
nutrient cycles 113–15, 125

obesity 206
oesophagus 183–4, 189
oestrogen 139, 140, 143, 155
    menopause 175
    pregnancy 146, 148, 150–1
    puberty 168, 169, 170–1, 177
oocytes 137–41, 150, 154
opsin 252
oral rehydration therapy 190–1
order 59, 68
osmosis 189–90, 266
ova 141, 154
ovarian follicles 138–9, 143, 150
ovaries 135–6, 137–8, 139–40, 171

oviducts *see* fallopian tubes
ovulation 138, 139, 140, 154, 178
oxygen transport 214–17, 222
oxyhaemoglobin 215, 222
oxytocin 150, 151, 153

Pacinian corpuscles 247–8, 259
pancreas 183–4, 187–8, 194, 270–1
parasites 182
Parkinson's disease 226
partial pressure 215–17, 219–20, 222, 223
penguins 31
pepsin 187, 188, 192
percentage saturation 215–17, 222, 223
peristalsis 238
pH 182, 184, 221–2
phenotype 13, 14, 26, 37, 49
phosphocreatine 258, 260
photolysis 97
photons 248
photosynthesis 95–9, 107, 115, 116
phylum 59, 68
physical activity ratio 202
pinocytosis 146, 147, 191
pioneer species 84, 88
pituitary gland
   birth 150
   development 163, 170, 171, 177
   lactation 153
   menopause 175
   reproductive cycle 139–40, 143
placenta 143, 145–7, 150–2, 154
Plantae 60, 65–6, 68
polarisation, membranes 229, 252
pollution 81–3, 122–3
polygenic inheritance 40, 49
polymorphism 47
polypeptides 187–8, 197
polyploidy 55, 56–7, 67
polysaccharides 183, 184, 187–8, 195
population genetics 40–3
populations 74, 75–9, 88
positive feedback 279
potassium ions 228–33, 240, 252
potential difference 229, 252
pregnancy 148–54, 203, 206, 208
primary consumers 104, 115, 119
primary succession 86
producers 104
progesterone 139, 140, 155
   menopause 175
   pregnancy 146, 148, 150–1, 153
   puberty 169, 170–1, 177
Prokaryotae 60, 61, 68
prolactin 151, 153–4
prophase 9, 10, 136
proteases 187
proteins
   absorption 191
   collagen 171–2
   diet 194–5, 197, 208
   digestion 183, 187–8, 189
   farm animals 208–10
   puberty 168
   requirements 199, 203–4, 207
   role 195

Protoctista 60, 62, 68
puberty 134, 137, 163, 168–71, 177, 178
Punnett diamond 16

quadrats 76–7, 88

random fertilisation 37
random sampling 34, 49, 76, 88
receptors 246–7, 259
   eye 248–52
   neurones 227, 237
   Pacinian corpuscles 247–8
   temperature 275
red blood cells 214, 218, 220–2
reflexes 193, 236–8, 240
refraction 248, 250, 259
refractory period 231, 233, 240
relay neurones 227, 236–8, 240, 250, 251, 259–60
reproduction 132–54
reproductive isolation 57, 58, 68
respiration 99–104, 107, 115–16, 270, 280
respiratory quotient (RQ) 102–3
resting potential 228–30, 231–2, 240, 247
retina 235, 246, 248–50, 259–60
rhodopsin 252, 260
rod cells 231, 235, 246, 250–2, 259–60
RQ *see* respiratory quotient

salivary amylase 184, 188
saltatory conduction 233
sample size 75
sarcomeres 253–4, 255–6
sarcoplasmic reticulum 253, 255, 257
Schwann cells 227, 228, 233
secondary consumers 104, 115, 119
secondary succession 86
selection 43–8
seminiferous tubules 133–5
sensory neurones 227, 236–7, 240, 246, 248, 259
Sertoli cells 135
sex hormones 168–71, 175, 177
sex-linked inheritance 21–3
shunt vessels 276
sickle-cell anaemia 47–8, 49, 218–19
significance 24
simple reflexes 236, 237, 240
skeletal muscle 253–9, 260
skin 171–2, 277–8
sliding filament theory 255–7
slow muscle 258–9, 260, 261
small intestine 183–5, 187–8, 189–92, 193–4, 207
sodium ions 189–91, 228–33, 240, 241, 247, 252
sodium–potassium pump 228–9, 230, 232
somatotropin *see* growth hormone
speciation 55–9, 67
species 54–5, 59, 67, 68
sperm 8, 132, 133–5, 137, 140–1, 154
spinal cord 236–8, 240
spindle fibres 9, 10
sporophytes 65–6, 68–9

stabilising selection 44
standard deviation 35–6, 49
standard error 36, 49
standing height 162, 164–5, 177
starch 183, 184–5, 188, 195
stimuli 230–1, 236, 246–8
stomach 183–4, 188, 194
stroke volume 149, 173, 177
succession 83–7, 88
summation 231, 235, 252, 260
supine length 164, 177
surface area 147, 189, 201, 208
Svedberg unit 61
sympatric speciation 58–9, 68
synapses 234–6, 237, 240
   ageing 174
   eye 251, 252
   receptors 246, 248
synaptic knobs 227, 228, 234, 240

T tubules 253, 255
taxa 59, 68
telomeres 176
telophase 9, 10
temperature homeostasis 265–6, 273–9, 280
test crosses 17
testes 132, 133, 135, 169, 171
testosterone 168, 169, 171, 177
tetraploidy 56
thermal balance 149–50, 154
threshold stimulus 231, 234, 240, 252, 260
thylakoids 96
thyroxine 170–1, 177
toxins 235–6, 241–2
transamination 198
transects 77, 88
triglycerides 183, 188, 192
trophoblast 142–3
trypsin 187–8
Turner's syndrome 11

umbilical cord 145, 146, 150
uterus 135–6, 140, 154
   pregnancy 142, 145, 148, 150–1

variation 31–6, 37–40, 49, 55, 67
vasoconstriction 276
vasodilation 276
vegetarians 207, 208
ventral root 236, 237
villi 142–3, 146, 147, 185, 189
visual acuity 252, 260
vitamins 194–5, 199, 206, 208
voltage-sensitive gates 228–32, 233, 234, 241

water potential 185, 189–91, 266
weight 164
wheat 56–7

xanthophyll 20

yeast 64

zygotes 142, 154